AI Ethics and Governance

Zhiyi Liu · Yejie Zheng

AI Ethics and Governance

Black Mirror and Order

 Springer

Zhiyi Liu
Computational Law and AI Ethics Research
Center of Shanghai Jiaotong University
Shanghai, China

Yejie Zheng
Computational Law and AI Ethics Research
Center of Shanghai Jiaotong University
Shanghai, China

ISBN 978-981-19-2530-6 ISBN 978-981-19-2531-3 (eBook)
https://doi.org/10.1007/978-981-19-2531-3

© The Editor(s) (if applicable) and The Author(s), under exclusive license to Springer Nature Singapore Pte Ltd. 2022
This work is subject to copyright. All rights are solely and exclusively licensed by the Publisher, whether the whole or part of the material is concerned, specifically the rights of reprinting, reuse of illustrations, recitation, broadcasting, reproduction on microfilms or in any other physical way, and transmission or information storage and retrieval, electronic adaptation, computer software, or by similar or dissimilar methodology now known or hereafter developed.
The use of general descriptive names, registered names, trademarks, service marks, etc. in this publication does not imply, even in the absence of a specific statement, that such names are exempt from the relevant protective laws and regulations and therefore free for general use.
The publisher, the authors and the editors are safe to assume that the advice and information in this book are believed to be true and accurate at the date of publication. Neither the publisher nor the authors or the editors give a warranty, expressed or implied, with respect to the material contained herein or for any errors or omissions that may have been made. The publisher remains neutral with regard to jurisdictional claims in published maps and institutional affiliations.

This Springer imprint is published by the registered company Springer Nature Singapore Pte Ltd.
The registered company address is: 152 Beach Road, #21-01/04 Gateway East, Singapore 189721, Singapore

Preface: Metaphor and Future of Humanism in the Wave of Science and Technology

Look in the mirror, and you will get a clear understanding of yourself.

Future Fantasy

In 1957, *Popular Mechanics*, a renowned American tech magazine, prudently proposed that roads in America would become pneumatic pipes by the early 2020s.

"Letters will be delivered by rockets. A letter can be sent from New York to California in a few hours." U.S. The Postmaster General said delightedly in 1959.

Voyage to the Prehistoric Planet, an American science fiction movie, was released in 1965. In the movie, after establishing a permanent base on the moon, astronauts set about to travel tens of millions of kilometers to explore Venus and discovered multiple exotic animals and plants on the surface of the planet, including dinosaurs that had been extinct on Earth.

In 1997, *WIRED* magazine made a similar prediction; that is, humans would be able to land on Mars approximately 2020.

2020 has arrived now, and various unexpected situations have been unfolding in an unrealistic way. The Australian mountain fire in January killed 1.2 billion animals, and the African locust plague and swine fever in February starved 19 million people. Since March, COVID-19 has caused over 10 million cases of pneumonia. In June, China's floods hit 24 provinces. However, in previous film and television works, 2020 sounded like an extremely distant year.

In the science fiction world created in film and television works, 2020 is a very special year. By now, human beings should have been able to easily travel to wherever they like and enjoy a high level of scientific and technological civilization. According to the incomplete statistics on Wikipedia, there are at least 97 films with timelines relating to 2020. In the imagination of these movies, cars in *Blade Runner* can drive through skyscrapers in cities; human beings in *Johnny Mnemonic* can directly transmit data with their brains without saying anything as long as they put on a type of helmet; Metalbeast, who is transformed by scientists into a part man, part machine cyborg in *Robocop*, works to guard the peace of Detroit; and the robots created by humans in *The Terminator* even manage to take over the world.

However, the world now has not become cyberpunk as in science fiction movies. Cars still cannot drive in the air, and AI cannot yet replace human beings. "Brainternet" has not happened, and it is impossible to buy an air ticket to Mars. Too many things have taken place without warning in 2020.

Although wild imaginations in various movies and television works as well as science fiction have not come true by now. Driven by science and technology, the greatest productivity today, the era of man-machine coexistence has come to some extent. The two industrial revolutions have led to a giant leap in human productivity. The fact that the knowledge and wealth being created during the past hundred years have outnumbered the combined achievements gained over the past several thousand years has made mankind unprecedentedly ambitious. Practical technologies such as robots, AI and cloud computing, which have become increasingly integrated into people's daily lives, it is reasonable for human beings to expect a more automated life.

Marx once said "the differences of all kinds of economic times, do not lie in what is produced, but how to produce, with what means of labor to produce". When technology has become a basic way of thinking and living, the prelude to the era of intelligent economy has opened. We need to realize that in the digital society, even if human beings dominate, the relationship between humans and technology also turns out to be a dynamically changing game-playing process. Different types of intellectual products, including driverless cars, face recognition and smart pills, educational robots, etc., are commonly seen in everyday life. These intelligent crystals represent unparalleled achievements on the combination of cognitive reason and practical reason. Especially since 2016, AI has outperformed human beings repeatedly, demonstrating the strong momentum of technologic rationality.

Being heavily influenced by traditional Western philosophy typified by Socrates, Plato and Hegel, enlightenment rationality becomes the gene and basic color of this rationalistic and revealing digital society. However, the stubborn pursuit of rationality and blind pride will inevitably make rationality originally serving as purpose change qualitatively, namely, utilitarian rationality has coercively blinded modern people to the opposite of a happy life.

Actually, the "digital society" itself implies contradictions. It is a combination of "society full of intelligent machines" and "human society". Since machines and human beings are completely different in nature and attributes, their integration is bound to be an extremely complicated process full of contradictions. Such contradictions have appeared in the early days. Repeatedly incredible victories won by machines, man's overreliance on machines, accidental injuries "conducted" by machines, and even horrific scenes displayed in science fiction movies about how machines rule and enslave humans have made the relationship between man and machine go far beyond master and servant during its constant upgrades. AI was originally a great invention for us to lead a better and more rational life, but technological rationality has deteriorated into excuses to satisfy human desires in multiple triumphs over human performance. Rationality is losing its nature in the cage of desire, while desires are behaving unscrupulously under the guise of civilization.

As mentioned above, technology has brought unprecedented power to human beings in science fiction movies, but they also show the "dark side" amplified by science and technology: abnormal human desires are exposed in front of super AI, and people are constantly suffering from their boundless avarice. The beautiful robot in *Ex Machina* is created for human carnal desires. In *AI*, people try to avoid facing death by resurrecting their dead relatives with AI technology. The hero in *Her* falls in love with his virtual girlfriend to escape normal human interaction in reality. Obviously, old tough robots can no longer satisfy the numerous demands of human beings, and AI products are being created on a large scale to stimulate both the physical and spiritual desires of humans for various purposes, including commerce, politics, and war.

It can be seen that while promoting social progress, science and technology may intensify the alienation and distortion of human nature, which in turn will further speed up technological advances and cause more negative effects. With more advanced and intelligent AI products coming down the pike, what they can do is just filling the gap of desire temporarily instead of building a harmonious ethical order between human and machine world. The protagonist of *Her* is addicted to the virtual intelligent world and finds it difficult to adapt to life in the real world. The modern version of Allegory of the Cave suggests that it is not the illusion in the dark but the light of reality that makes us afraid at present. The super AI in the movie is not just a fantasy; it indicates that modern people are no longer satisfied with cold technological products but want to invent a human-like plaything to meet their selfish desires through omniscient role play. By doing so, they are free from the responsibility of reality. Does this mean that technology is nourishing and encouraging the selfish gene of humankind?

Moreover, the relationship between AI and humankind is a topic interconnected with the modernity of human history, as Durkheim stated: "Modern times derive from the past and finally become the past although they always position themselves against the past". After entering the postmodern period, it is still difficult for us to find a similar relationship pattern in history, especially how to get along with AI in which they lack experience. The imaginations for the future, whether optimistic or pessimistic, all reflect different assumptions and expectations about the good and evil nature of human beings.

Marx's concept of alienation is based on the assumption that all men are "born" to be kind but later corrupted by society. In contrast, Durkheim's "anomie concept" holds that men are "born" to be a stubborn organism whose conceit must be strictly restrained by society. The former view is considered close to that of Rousseau and the latter to that of Hobbes, but all of them are related to the discussion on human nature brought about by modernity.

In this sense, the imaginations for the future are actually rules and paradigm propositions for humans to enter a world of "human-computer symbiosis", rather than just about the imaginations of science fiction on social development.

Metaphor of Black Mirror

As an important product of the era of human-computer symbiosis, ubiquitous electronic devices are penetrating into people's everyday lives. On the one hand, they draw an insurmountable boundary between the so-called "real" world and the "virtual" world; on the other hand, they are the only medium for the two to communicate with each other. The screen will bring us endless information once it is opened, and people will be immersed in the illusion of sound and light it provides without being aware of its border. The black mirror's materiality is exposed to us only when it is closed. The mirror carries an unfathomable force of technological ethics. As Nietzsche said, "When you look long into an abyss, the abyss looks into you."

China and the West hold different perspectives on the understanding of multiple metaphors of mirrors. Le Daiyun discussed the internal differences between Chinese and Western cultures by taking "mirror" as the clue of her article "mirror metaphor in Chinese and Western Poetics". She pointed out that westerners focus more on the reflection function of "mirror". Stendhal, Goethe, Shelley, and other writers compared literature to a mirror to highlight the reflection of literature on reality. Relatively speaking, Chinese culture prefers to use idioms such as Jinghua Shuiyue and Yijing Zhaojing, which literally means the reflection of a flower in the mirror and of the moon in the water and using the images in the mirror to realize introspection, respectively. Deeply influenced by Buddhism and Taoism, Chinese culture places more emphasis on the features of emptiness and inclusiveness of mirrors. It is often used to describe the inner feelings of humans. The most representative image should be the alternate name of *The Dream of Red Mansions*, "*Fengyue Baojian*" (mirror in breeze and moonlight).

When looking at other cultures, we find that mirrors have more complex and diverse cultural connotations. For example, buildings such as Mausoleum of King of the Light in Shirazi, a city in southwest Iran, and Golestan Palace in Tehran, the capital, are mainly decorated with fragmented mirrors, which form the dense geometric ornamental motifs in the inner wall of the building. Leaping lights crack into innumerable tiny light spots through the reflection of the fragmented mirrors, making up a resplendent and flexible view with a sense of holiness. More importantly, this decorative way takes advantage of the properties of mirrors while circumventing religious taboos—Islam forbids the use of images of people or any creature with eyes for decorative purposes, and countless small and interlacing mirror shards just avoid any recognizable figure.

After completing his famous fairy tale *Alice in Wonderland*, Lewis Carroll, an English writer, created another *writer through the Looking-Glass and What Alice Found* in 1871. This time Alice does not fall into the rabbit hole but enters a mirror world through the mirror at home. As an intermediary, the mirror gives this dreamland some rules contrary to reality. For instance, the closer you want to get to a distant mountain, the farther away you are from it; you have to turn a cake first if you want to cut it and so on. Meanwhile, the mirror also divides the world into two parts, a surface world and its opposite, an absurd world filled with the satire and parody of the former one. The opposite of the surface phenomenon world is the absurd world of satire and parody. It can be said that the introduction of "mirror" adds a touch of allegorical dimension to this fairy tale and interrupts its original narrative rhythm. At the end of the story, the author describes how Alice wakes up and excitedly looks for the real counterparts of the characters appearing in her dream, such as the Red Queen, the White Queen and the Humpty and Dumpty, at great length. In the tale, the mirror also separates the phenomenal world from the mirror world, and the latter implies the reality of the former in the opposite way.

In contrast, the cultural metaphor of the "mirror" in the British drama "Black Mirror" is much more complex. Differing from the metaphors of the above classical mirror, the mirror here is no longer the dividing mark between real and illusory binary imagination. Instead, it allows the mirror world to occupy and replace the real world by reducing its presence. In the play, the traditional metaphor of mirror separating the two worlds becomes something like a two-way transparent membrane that allows the fusion and transition of subjects in the mirror and their realistic counterparts. Such imagination keenly grasps the features of the digital mirror and displays them in an ironic and absurd way. Undoubtedly, the "mirror metaphor" in culture has something to do with the construction of subjects, and the *Black Mirror* also uses mirrors to remind us of the dilemma faced by the increasingly blurred and broken subjects in the technical age.

On the one hand, *Black Mirror* profoundly reveals the conditions of people who live in a highly mediated posthuman world. The mediated existence mode subverts the perfect fantasy of human beings as rational subjects as well as their dominant position at the technical level. Lacan's "mirror stage" theory notes that self-concept is built entirely through identification with others (mirror image), which is external. The "self" is deceived from the very beginning by the mirror image, that is to say, the mirror images we see are an imagined "ideal self", and it is the product of our imagination. This makes our desires become others' desires, and we turn into the "silent majority". People hide behind the black mirror, unleashing their dark natures. This seems to be driven by the nature and desire of selves, but no one can escape from being influenced and utilized by the media manipulator. From this point, media, with great symbolizing capability, becomes the "the big Other" proposed by Lacan. As such, the media has completed the "castration" of human beings and pushed the unipolarity of the human desire structure. A medium catalyzed by technology can either bring hope to mankind or just destroy everything like fire.

On the other hand, *Black Mirror* further deepens the thoughts of alienation via the demonstration of a large number of possible cyborg forms. Cyborg is a combination of living body and media technology. Cyborg in the TV drama can be divided into two types: physical cyborg and cyberspace agent. The former exists in reality, while the latter exists in cyberspace. They interpenetrate to each other. What challenge does Cyborg bring us? It makes us think about whether humans have essence. Do "I" still exist if my body is completely replaced by others (machines)? Is "consciousness" reliable at that moment?

It can be argued that this high-profile cultural work is not only the product of technically pessimistic anxiety, and its greatest value actually lies in its scientific and technological reflection of the coexistence of criticism and reflexivity. As a well-regarded episode of Black Mirror, *Fifteen Million Merits* has built a "beautiful new world" in which all people have to amuse themselves to death. Humans in this world are surrounded by a diverse variety of scientific and technological electronic screens every day, and all they can do is keep riding on the indoor sports bike to provide power for all electronic screens and earn "merit" for themselves. Those whose merits reach 15 million can participate in a TV program called Hot Shot (an acrimonious version of *Britain's Got Talent*). Managing to be scouted by the judges on the program through participation is the only way a person can get rid of the fate of pedaling every day and be forced to watch assorted entertainment and pornographic shows. To help his beloved girl Abi realize her music dream, Bing, a black man donates his merits to Abi and accompanies her on the stage of *Hot Shot*. It is on this stage that the song about "love" is sounded for the first time: "You can blame me, try to shame me. However, anyone, who knows what love is, will understand." Ironically, the judges neither understand "love" nor care about Abi's dream. Instead, they even suggest that she seek development on pornographic channels with a tone of menace.

According to the scenarios of Black Mirror, Abi finally accepts the suggestion that "love" is excluded from the "world", and only endless "sex" are being left in this system, which is similar to the story written by Huxley 80 years ago: sex can be enjoyed to one's heart content, but love is not a necessity. Indignant Bing makes every effort to earn back all the merits that need to return to the stage. After the performance, he holds a fragment of the broken electronic screen against his throat and directs a torrent of abuse at the judges present: "And all you see up here, its not people. You don't see people up here, it's all fodder... We don't know any better. All we know is fake fodder and buying shift. That's how we speak to each other, how we express ourselves is buying shift. The peak of our dreams is a new app for our Dopple; it doesn't exist! We ride day in and day out going where? All tiny cells and tiny screens and bigger cells and bigger screens."

It can be said that this passionate speech has basically covered multiple topics involved in this "unfettered drama", which includes technology prediction/fable, consumer society, media control, technical training, and human alienation. Of course, the "beautiful new world" in the teleplay exposes not only unavoidable problems caused by the "impact" brought by the technology but also all sorts of crux in modern society with the help of the "mirror" of science and technology. To what extent can an AI product implanted memory chip replace its real human counterpart (Be Right Back)? Can the cloned virtual consciousness be entitled to human rights (White Christmas)? Waldo, a virtual character who is well versed in the verbal tricks of the mass media in *The Waldo Moment*, finally becomes popular in political circles, which forms an interesting contrast with Trump's win. In addition, *White Bear* and *Nosedive* displays a series of current social behaviors, including taking photos, onlooking, scoring, giving likes, etc., through the power of science and technology.

This is the "Black Mirror"—when the light on the screen fades, the dark panel finally reveals the truth being covered, leaving "the world becoming ugly first, and then going out". Technology here demonstrates its desolate side. However, what cannot be ignored here is the mirror reflecting interactive images. It carries reflexivity as a metaphor. Once human beings enter the virtual world, their subjectivities are deconstructed first and then reconstructed into another form in a new encoding mode. When we reconstruct retrospectively, we obtain a distorted "I". This is the huge gap between the "second self" (i.e., cyberspace agent) and the real self. More seriously, the "second self" will cause harm to the real self, and the Internet facilitates the substitution of the second self with the real self. However, after that, the second self cannot possess human emotions such as love and hatred. Then, comes a reflexive paradox, which makes post humanism move toward its opposite, leaving post humanism hovering like a ghost in the spiral structures of "spirit-body-spirit" and "humanism-post humanism-humanism".

The black mirror metaphor is the reflection of human nature amid the carnival brought by the waves of technology, which is also the key for us to understand the whole book and promote the exploration of a new paradigm of social governance in reflection.

Ethics and Governance

Returning to the digital society itself, let us reflect on what is the original intention for inventing and manufacturing AI. Because we want to pursue a harmonious and beautiful life, a civilized society in which human beings can be friendly coexist with intelligent machines. Such a society is defined by the following three points:

First, the digital society is a real world instead of a virtual space. If we blindly pursue a technical and rational lifestyle and become lost in the trap of excessive consumption, we will forget the true colors of life and unconsciously live an "unreal life".

Since the twentieth century, phenomenology has unfurled the banner of "attending to the things themselves". People's reflection and criticism on scientific and technological rationality and instrumental rationality have never stopped. In a sense, the existence of the digital society is exactly shown through such ubiquitous digital quantification. Tasks such as monitoring a manhole cover, recording the track of a car, and even reading someone's mind can be comprehensively demonstrated by almost any AI entity via succinct and infallible numbers.

However, real human society is quite a different story. It is a life world. Human beings, the source of strength in the living world, follow the rule that existence precedes essence. For AI, the situation changes. People are active, generative, creative, transcendent and vibrant, and the corresponding life world is ever-changing, rich and higher than the scientific world. Changing our ways of life and thinking by copying the quantitative approaches in the scientific world down to the last detail is bound to destroy people's subjectivity and initiative, leaving a happy life impossible. Only by taking the life world as the starting point can we more reasonably explain and understand the scientific world represented by the digital society. The application of AI is not for self-protection and continuation of our species but creates a life helper to liberate people from daily chores, give better play to the creativity in humans, and enable them to have more leisure to enjoy life as well as the convenience of the digital society.

Second, the digital society is free and comprehensive and not unique to the elite. In other words, digital society needs to liberate not a small number of social elites but most ordinary people. AI may be a tool differentiating social and cultural, but on the other side of the struggle, the contradictions among people may also undergo subtle changes in the digital society. Especially when these contradictions become increasingly sharp, the necessity of friendly communication among people is further highlighted. After all, human beings are not AI. Humans' essence must be formed in social practice, and the meaning of life also needs to be constructed through the perceptions of subject experience. Only by getting along with people (not machines) and widely experiencing society can human beings live a more cheerful and unforgettable life and gain the most cordial feelings of happiness in the most natural state. We fully expect a more civilized digital society in which friendly relationships among people, social fairness and justice are more valued.

Third, the digital society features happiness rather than indulgence. Various novel AI products, such as household robots, helpers for the elderly, child care robots and pet robots, are constantly stimulating our consumption desire. They free us from physical fatigue to some extent and provide us with all sorts of sensory pleasure.

People often mistake these pleasure sensations as Epicurean's principle of joy, which emphasizes not only physical relief but also inner peace. Pure physical pleasure is a kind of negative happiness, but those who can overcome desires within them can acquire true pleasure. In the digital society, real happiness never stays only at the sensory level and certainly comes from the enrichment and tranquillity of human spirit. People can embrace their limitations and imperfections as human beings and refuse to be enslaved by desires, generating negative pleasure. Men should live a rational life without disguise, that is, a human life rather than a god's life. We can truly appreciate the beauty of the digital society only if we are able to control our desires and return to human rationality.

As we are still far from the ideal digital society, the technology industry is used to looking forward and paying more attention to product innovation, technology application and industrial realization, which is a good thing for the industry. When besuited senior executives and dedicated programmers in the office building are pressing ahead with the above work in unison, few are willing to spend time carefully spotting and predicting problems at the turning point in a rear-view mirror or even accepting a beneficial practice. This is the challenge we are facing. In other words, we can use the well-organized information in the past to grasp the growing trend of science and technology and confront the increasingly severe social and ethical problems head-on in the field, which is significant to any technology company on the "highway".

In recent years, some quality companies have started to pay more attention to the rear-view mirror because they keep constantly adjusting their development directions and speeds. In fact, a number of large technology companies have focused attention on the ethical problems of AI and took action as early as 2017. Since 2011, Microsoft,

IBM, and the University of Toronto have realized that it is extremely necessary to overcome bias and accomplish fairness in AI. This is because from the perspective of technology, most algorithm models in AI work like black boxes. Usually, researchers only know the input, output and quality, but they cannot explain the internal working mechanism of these algorithms. If AI researchers adopt biased data sets or add their own prejudices when adjusting parameters, unfair AI outputs may occur.

To solve this problem, MSRA created a project named FATE (Fairness Accountability Transparency and Ethics in AI) in 2014 to increase the fairness, reliability, and transparency of AI. The project group organizes a symposium called FAT in ML every year and invites many senior technical experts in AI to share new research progress. Deepmind, a British company acquired by Google, also founded a project named Deepmind & Society in 2017 and put forward two goals. One is to assist technicians in practising ethics, and the other is to help society predict and guide the influence of AI for the benefit of all.

For technology companies, AI is far from a panacea but is a very risky technology. Thus, they need to be capable of "looking forward and rethinking the past" to be competitive in the industry. Before the "black box" of AI can be clearly understood and fully explained, a good suggestion is to formulate the basic laws for AI. Of course, as the impact of AI has gone far beyond the average organization, countries worldwide are laying down AI governance principles and proposing corresponding strategies suitable for local AI development. The most representative AI governance law should be the EU's ecological governance law based on AI ethics.

As early as January 2015, JURI decided to form a working group to study legal issues related to the development of robots and AI. In May 2016, JURI issued the *Draft Report with Recommendations to the Commission on Civil Law Rules on Robotics*, calling on the European Commission to assess the impact of AI. In January 2017, JURI formally put forward a wide range of suggestions on robot civil legislation and the proposal to formulate "Charter on Robotics". In May, the European Economic and Social Committee (EESC) released a statement on AI and pointed out an initiative to establish AI ethical norms and a standard system for AI monitoring and certification in response to the opportunities and challenges brought by AI in eleven fields, including ethics, security, and privacy. In October of the same year, the European Council proposed that the EU should have a sense of urgency to cope with the new trend of AI and guarantee high-level data protection, digital rights and the formulation of relevant ethical standards. The organization also invited the European Commission to give solutions to deal with the new trend of AI in early 2018. To solve the ethical problems caused by the development and application of AI, the EU has set AI ethics and governance as the key to future legislative work. Undoubtedly, the EU is taking the lead in promoting localization ethics to governance.

On the basis of the current situation, we suppose that AI governance is an indispensable proposition for organizations (countries, communities, enterprises, households, etc.) entering modern society. Therefore, it is necessary to write an academic monograph ranging from ethics to governance to explore the systematic logic of human beings from social ethical reflection to ecological governance, which can be regarded as an innovative attempt from technical ethics to technical governance. It can be seen that since we are experiencing the fourth scientific and technological revolution, the ethical confusion resulting from technical development has become a new normal when we talk about big data, AI, gene pairs, and reproductive cloning. This process involves an obvious paradox. As a means for human beings to expand the world and nature, technology also extends to human beings through rapid self-extension. In short, human beings have eventually become a part of humanized nature. The explosive development of modern science and technology has driven us to review the definition of life, especially the relationship between humans, the "subject", and other species. At the moment, the humans who seem to have "controlled" everything through technology are getting confused. We raise the eternal propositions again as our ancestors did thousands of years ago. What is the essence of life? What is the core difference in value between man and machine? Only when we determine these propositions can we have a clear thought when we think about the ethics of AI.

The results of the digital technology and AI revolution cannot be predicted because history can never tell us what will happen in the future. Severe direct threat, long-term structural imbalance, and tense situations have triggered a heated debate on AI ethics and governance throughout the world. In the debate, the term "ethics" is often used to describe legitimate concerns about the potential destruction of AI. The most controversial part in the discussion on AI ethics and governance is all sorts of framework definitions of AI principles worldwide, which are mainly put forward by large Internet platforms, multinational corporations, international nongovernmental organizations, and governments around the world.

Although there are some subtle and key differences among these ethical principles, each of them emphasizes that the future AI should be safe, interpretable, fair and reliable, and it should benefit the whole society. All these definitions seem to reach an international consensus; that is, AI should serve the welfare of mankind, be people-centered, responsible and trustworthy, always let human beings take the initiative and be kept under supervision.

However, these positive principle frameworks confirm that today's ethics and governance capabilities are insufficient to prevent or mitigate the destructive power of AI, which has a fatal global impact and historical significance. Almost all frameworks analyzed the risks of AI in a narrow sense without considering the relationship between the dual attributes of technology and actual social, political, economic and international affairs. These frameworks ignored that the historical trajectories of society, politics, economics, and international affairs are likely to be strengthened rather than changed by AI. AI will make an increasing number of independent decisions, but it cannot escape human control and obtain complete autonomy in the short term. We can never rely on AI to assume the role of an excellent, super beneficial, and people-centered compass to guide humans to live a generally fair and dignified life. Although AI principles are easy to establish, given the complexity and uncertainty of the risks of AI, it will be more difficult to implement AI principles according to the newly formed governance methods.

To solve the problems above, we need to have an interdisciplinary knowledge system. In this book, we will discuss not only humanities frameworks such as ethics, political economics, philosophy, and sociology involved in technical ethics but also science and engineering frameworks such as computer science, biology, AI, etc. We believe that only by breaking the boundaries of disciplines and expanding our knowledge to the cross fields of corresponding disciplines can we generate helpful ideas for complex problem-solving in a world of uncertainty. Fortunately, we have already begun to make rules for AI before it acquires real intelligence, and that is also the reason why we decided to write this book. We are expecting to expand and construct humanism in the era of AI.

For today's people, breaking through the early ideas of the Enlightenment era and constantly exploring human beings, the external world and the relationship between them is an essential and positive choice. Such exploration is multidirectional, including not only the continuous research and invention on life science, AI, and other science and technology but also research on the solutions to these existing or upcoming phenomena and even prevention measures. The latter requires collaborative cooperation among humanities, social sciences, and natural sciences.

If you review the past billions of years of the planet, Homo sapiens created human civilization in only 7,000 years and became the most important force on the planet that has got rid of the constraints of nature and been able to decide on its own development. What is the difference between humans and other animals that makes them a special species? Among multiple theories, the "culture driven theory" based on information transmission efficiency proposed by Kevin Leyland, a British scholar, has been well supported. The theory holds that the difference between human beings and other animals lies in the former having rich oral or symbolic language, which can help them learn more from culture and inherit more than the animals. The amount of cultural reserves of a species and the duration of its cultural characteristics in the population will increase exponentially with the rise in fidelity. Among the existing organisms, humans are the only animals that have broken through the threshold of cultural change to form the so-called "coevolution of gene and culture" process. In other words, the reason why large-scale cooperation occurs in human society is that we have social learning and educating abilities with which we can not only copy the cooperation mechanisms in other organisms but also generate new ones of our own, thus forming an unparalleled population.

Hence, the foothold of our research on AI ethics is humanism; that is, we need to safeguard people's central dominant position through human-centered cultural value inheritance and the construction of a social knowledge system. We can see that the humanistic spirit in the era of AI aims to continuously promote and plan human development and evolution where possible and improve human beings with the help of rapidly changing technology. If human nature cannot develop, gradual elimination and extinction rather than just suspension and standstill will be their final destinies. Humanism in the era of AI covers the following self-confidence and spirit: the development and evolution of human beings have shown the trends of unity of culture as well as the unity of physics, which forms the basis for people's positive views on dealing with human prospects.

Over the past few years, scientists have proposed using "humans" as a calender unit of the earth. The proposal of this geological concept is based on vital observation and research data, which represent the current situation brought to modern humans by the 7,000-year history of human civilization. Before the Anthropocene, the power to change the earth's ecology and climate on a large scale came from the earth itself, while natural selection has been gradually replaced by artificial design since humans emerged. Moreover, life forms have been extended from the organic field to the inorganic field, and artificial design has also been put into nature to shape the environment. Most importantly, human beings are using the power of technology to influence their own evolutionary direction, including bioengineering, cyborg engineering, nonorganic bioengineering, etc. All the problems in the fields above are closely related to the ethics of AI. Essentially, it is something about self-cognition and technical ethics, which is also the basis of the book that we want the readers to understand. What is the essence of the ethical problems of AI? How can a reasonable order be established to achieve man-machine symbiosis? We will elaborate on these propositions in the following twelve chapters. The content of the book, ranging from the modernity of social development to the discussion on the social risk of digintelligence, from AI medical discussion to in-depth synthesis technology, and from computational law to man-machine cogovernance, is expected to provide enough space for readers to understand AI ethics and governance that might matter the survival of human beings. We are glad to welcome all readers to embark on a "fantasy thinking journey" about the future of mankind with us.

Shanghai, China	Zhiyi Liu
Shanghai, China	Yejie Zheng

Contents

1	**New Problems in the Era of AI**	1
	1.1 What Makes a Human Being a Human Being? AI and Gene Editing	2
	1.2 Man-Machine Symbiosis: Challenges and Problems in the Near Future	7
	1.3 Data Ethics: Political Economy in the New Oil Age	12
2	**Ethical Enlightenment in the Age of Intelligence**	17
	2.1 Development of Human Consciousness and Machine Consciousness	19
	2.2 Looking at AI from the Differences Between the East and the West	22
	2.3 Ethics and Moral Responsibilities of AI	27
3	**Digintelligence Risk Society is Around the Corner**	33
	3.1 AI Ethics in Digintelligence Risk Society	35
	3.2 Will Winter Come?	39
	3.3 Data Privacy and Social Responsibility	42
4	**AI Medical Treatment: Epidemic, Death and Love**	47
	4.1 Black Swan Incidents: Epidemic and Influenza	48
	4.2 Breakthrough and Ethical Principle of AI Medical Treatment	52
	4.3 Restart AI and Expand the Meaning Boundary of Life	57
5	**"Secret" Left by Turing—Privacy Computing**	63
	5.1 Ethical Turn of In-depth Learning: Theory of Materialization of Morality	64
	5.2 Fantasia in the Turing Era—From Encryption to Computing	69
	5.3 Secret Greatness: Creating Privacy Enhancing Technology	73

6	**AI and Robot: Darwin and Rebellious Machine**	79
	6.1 Paradigm Revolution of Cognitive Science: Taking Consciousness Research as an Example	80
	6.2 Human-Computer Symbiosis Under a "Brain-Computer Interface"	82
	6.3 Automatic Driving: "Survival of the Fittest"	87
7	**Virtual World Under AI: Augmented Reality and Deep Synthesis**	95
	7.1 AI+AR: A New World Under Augmented Reality	96
	7.2 New Risks of Deep Synthesis Technology	101
	7.3 Masked Virtual Digital Humans	105
8	**Start of the "Age of Exploration" of AI Governance**	111
	8.1 New Channel or Old Ticket	112
	8.2 Excellence, Trust and Reliability	116
	8.3 Human Versus AI: "Victory" of Anthropocentrism	120
9	**AI Legislation in Computational Society**	127
	9.1 Who Do We "Legislate" For?	128
	9.2 "What Makes a Machine Human"?	132
	9.3 The Foundation of Civilization Comes from "Dignity"	136
10	**New Rules and New Order in the Era of AI**	143
	10.1 Creating a Technological Contract Between Humans and Nature	144
	10.2 "New Rules" of Human-Computer Interaction	146
	10.3 "New Order" for the International Community	151
11	**How Can "We" Realize Cogovernance in the Postepidemic Era**	159
	11.1 Public Goods from a Global Perspective	160
	11.2 Governance and Innovation of Intelligent Society	164
	11.3 Truth of AI and Reflection of Black Mirror	169

Chapter 1
New Problems in the Era of AI

With the application of a range of technologies, such AI, big data, gene editing, augmented reality, blockchain, etc., the way we live has changed greatly. Science and technology have brought us a healthier and more comfortable life, liberating our hands to allow us to have more opportunities for cultural exploration and self value fulfillment. In particular, AI and other technologies have played a significant role in the global fight against the pandemic. It can be found that AI techniques such as medical image-aided diagnosis, UAV contactless service, detection and prevention of automation have gradually gained popularity in various scenarios. Meanwhile, China has made it clear that data have become a production factor in the era of a digital-driven economy. How to standardize and promote the use of data has become an essential topic for the development of AI.

The global outbreak we are facing is a fight head-to-head between humans and viruses. Will technology stand in the same boat with human beings when facing such "black swan" incidents? We need to face up to this fundamental problem here—the relation between technology and mankind. Where will technology, one of the most powerful tools of mankind, lead humans in the future? If technology is compared to a medicine, then what are the side effects of this medicine?

Although technology in itself is neither good nor bad, its feature may change when mixed with complicated human nature. A good tool may sometimes turn into dangerous "black technology". Criteria for good and bad rooted in social cultures can greatly influence the development of technology. Each technology today is marked by social cultures of humankind. Therefore, the analyses of technical ethics must be placed in the social and cultural background. In other words, technology itself does not involve any qualitative ethical judgment, but it creates specific conditions and space for the corresponding ethical judgments. In this way, human beings can discuss technology in a certain social culture. It cannot be simply explained by "technology neutrality theory", but it indicates that the image of human nature being reflected on the mirror of technology can directly affect the development of technology and the corresponding ethical choices.

© The Author(s), under exclusive license to Springer Nature Singapore Pte Ltd. 2022
Z. Liu and Y. Zheng, *AI Ethics and Governance*,
https://doi.org/10.1007/978-981-19-2531-3_1

Let us first look back on the background of technical ethics. For a long time, technology did not show its importance in philosophy and ethics until the period of industrial revolution. As technological modernization pushed human society into modernity, the significance of technological ethics is remarkably rising. The word "modern" did not exist until the 16th century. It means "now or present" in English. Moreover, "modern times" represent a new conception of history. History repeats itself again and again before modern society. After the industrial revolution and the Enlightenment, people began to shift their attention to the "current era", and then time became a linear and directional vector. This is the core background of "technical ethics". By then, people start to deny and break with the tradition, turn their eyes to the present and future, and further reinforce human creativity and subjectivity. This indicates that human civilization has moved into a modernization process. The process has also brought the problems of modernity at the same time.

The author will explore the ultimate problems brought by technology from the perspective of technical ethics in this chapter. These problems include but are not limited to the following: should we use gene editing technology to change the method of reproduction and evolution? Can cyborg help sustain the continuation of our species? Will AI control humans? How do we formulate data rules in the era of digintelligence… It is worth pointing out that we prefer to take a further look at these future technologies with certain "post-human" attributes, especially gene technology and AI technology, from the angle of digital economics rather than traditional economics.

1.1 What Makes a Human Being a Human Being? AI and Gene Editing

On November 26, 2018, He Jiankui, a former associate professor at the Southern University of Science and Technology, announced at a conference that a pair of gene editing twins was born in China in November. One of the twins' genes was modified to prevent HIV infection. If that worked, they would become the world's first gene editing case that is immune to HIV. He Jiankui's announcement drew widespread condemnation in the global scientific community. Many world gene research pioneers criticized that his research had departed medical ethics. Meanwhile, China's medical and scientific research regulatory authorities immediately filed the case for investigation. This "farce" resulted in the birth of a baby with a modified gene for the first time in the history of the world. The case is also commonly believed immoral by the scientific community and seen as a trample on human dignity and scientific spirit.

Technically, CRISPR gene editing adopted by He Jiankui received a great deal of attention, and the case started an ethical debate on human germline manipulation. However, the more central issue is that before CRISPR research can be safely transformed into treatment, scientists need better methods to avoid the potentially destructive target effect of this technology. In other words, the problem lies in the fact that DNA will either repair itself or mutate during the process after the CRISPR-Cas9 editing tool cuts double-stranded DNA.

Gene editing technology has completely broken the restrictions of traditional medical treatment. If not being used properly, the technology can completely subvert the existing landscape of human society in an uncontrollable way and may even turn into the source of human destruction. This is why the event has attracted much attention worldwide. We should be cautious about the development and application of gene editing technology. However, it is undeniable that compared with the strong influence of gene editing technology, traditional medical treatment actually provides limited help to human beings.

In essence, although the technology of traditional medicine is still advancing, the fact that the essence of "traditional medicine is just a kind of support for the self-healing of life" can never be changed. Thomas Szasz, an American doctor, once said, people mistook God's power for medical treatment when the influence of religion proved to be greater than that of science; people are pinning all their hope on medical treatment when the opposite happens." Indeed, a tacit understanding shared by all doctors is that it is the patient himself who truly cures his disease, and all medical treatment just plays a supporting role. That is, the self-healing ability of life is what truly counts. Medical support aims to win time and create conditions for self-healing, waiting for self-rehabilitation to finally play its role and triumph over illness.

Today's gene editing technology has changed the behavioral logic of traditional medicine and directly explored the means to improve human existence. Genetic manipulation of human embryos in test tubes may not only help prevent genetic diseases but also change heights, intelligence and other characteristics of humans. In the wave of gene editing technology, there is a strong chance that "super eugenics" will emerge in the future. Cross-use of gene manipulation, high-quality genes, cloning and some traditional methods will thoroughly subvert our understanding of "medical treatment" and directly alter the future of human society. Moreover, enhancing the sensitivity of human perceptual intuition and expanding the scope of human motion control through a brain-computer interface is also an approach to improve human beings.

The way of changing the nature of medical treatment by manipulating genes to "strengthen human conditions" has been ethically opposed. Because a large number of scholars believe that such a practice makes human beings lack "naturalness" like handicrafts, which is detrimental to human dignity. However, this excuse seems to be increasingly challenged, especially when we face unpredictable disasters and unhelpful modern medical procedures. At that time, we often turn to such technical solutions. The contradiction here lies in the fact that we choose to advocate human dignity in the face of various events. However, medical defects, diseases and the promotion of genetic ability are essentially the same thing, but it is difficult for human beings to treat them equally in terms of ethics and emotions. This means that the combination of AI and gene editing technology challenges the fundamental ethical proposition that "what makes a human being a human being?".

From the perspectives of economics and sociology, how to strengthen humans is a great challenge for us. This is principally because the technology may serve those affluent groups and families, resulting in permanent social inequality, namely, the expansion of "free eugenics". The rich who occupy a large amount of resources can modify their offspring by editing genes so that their offspring can gain an overwhelming advantage over the average person in intelligence, appearance, height and even life expectancy, thus forming a genetic aristocracy composed of wealthy people. By doing so, they can monopolize all resources. By then, compared with genetic aristocrats, ordinary people without any strength will be eliminated to the bottom of society and may even become slaves or die out, leaving hierarchical solidification worse and worse.

In some extreme cases, if it were possible to judge genes subjectively, the tragedy of Nazi genocide might be repeated. As described by the British writer Aldous Huxley, in his novel, "*Brave New* World", published in 1932, all infants are incubated in laboratories instead of through natural viviparity. The fates and conditions of these babies are directly predetermined before they are incubated. They are classified into five "castes" or social strata, including alpha (α), beta (β), gamma (γ), delta (δ), and epsilon (ε). The target of all conditions set is to make people content with their inescapable social destinies. Five "castes" are cultivated separately. Among them, alpha and beta, the most upper class, lead and control other classes; gamma, the ordinary class, is equivalent to civilians; delta and epsilon, the lowest class with low intelligence, can only does ordinary physical labor… If the situation continues going like this, the long-established social ethical relationship will be overturned, let alone social equity. Ultimately, this will result in the disintegration of human society.

In addition, from the perspectives of the technical level and human driving force, these strengthening technologies will bring us to the "post-human era". The term "posthumanism" appeared in contemporary social science in the mid-1990s, but its source can be traced back to at least the 1960s. The philosopher Foucault also mentioned in *The Order of Things* that "man is a recent invention within it… Man would be erased, like a face drawn in sand at the edge of the sea and is approaching its end." The sentence can be summarized as his declaration of the "death of man". In all the ancient myths of mankind, the difference between man and God is that man is immortal, and God is immortal. When man breaks through his biological limitations and obtains eternal life, man becomes God.

Posthumanism calls the new human transformed by technology posthuman and reflects on it from two research approaches. One is the vision of human perfection driven by new technologies based on the summary and outlook of the development and utilization of technology. The other is the inquiries into posthuman and its era from the positions of reflection and criticism. Although these two approaches have a general sequence, they both focus on the boundary and changes in the form of "human" and the corresponding multidimensional thinking in the context of the high development of technology.

In this sense, posthumanism can be regarded as the deconstruction of humanism, and it dispels the particularity of humans in nature from the level of body and species. There are three evolutionary models for posthuman: genetic engineering or asexual reproduction (such as cloning technology); technical cultivation or artificial cultivation; the third way is to create virtual subject and transform real subject with virtual technology, combining virtual world with real world, virtual person with real person, which are prominently presented by a man-machine combination-cyborg.

In *The Letter from Utopia*, the famous philosopher Nick Bostrom strongly advocates the benefits of the posthuman era and summarizes three characteristics of posthuman states:

> First, most people can fully control their sensory experience by virtual technologies such as brain computer interfaces. There will be no difference between simulated life and real life.
> Second, the majority of people no longer need to suffer from psychological pain, such as depression, fear and self-loathing. The problem here is that we cannot determine whether there is a natural link between these painful experiences and personal achievement and self-esteem and whether we can embrace the benefits of "man-made" happiness.
> Third, most people will have a much longer life expectancy than their natural life span. The age of longevity not only results in aging, delayed retirement and labor shortages but also makes tediousness and irresponsibility common.

The Letter from Utopia depicts the happy side posthuman brings to us and awakens exciting scenes. People can rethink the meaning of human beings in new ways, being away from old constraints. However, from a dialectical point of view, the "evolution" of mankind needs to be transformed from the inside to obtain a new cultural form. Otherwise, the individual will be eliminated by times. Especially after the increasing technicalization of human beings, the debates on whether "technology can replace human beings" will break out. The former will cause disembodied posthumanism, and the latter will embody posthumanism.

Let us return to real cases. In the past few years, driven by the power of capital, some commercial hospitals have been researching gene editing technology. Scholars such as He Jiankui are conducting gene experiments regardless of ethical risks. The driving force behind this is the game between capital and power. On the one hand, biological gene knowledge is capital, and the capitalization of knowledge brings power. Many overseas hospitals make profits through the commercialization of knowledge and pay universities in the form of patents. In other words, as the life gene codes of life bodies are all being transformed into income, human beings are slowly losing their dominant right to control their own genes. Actually, capital promotes the knowledge system of production and accelerates the deviation of the human central position, contributing to the distortion of ethics and values. On the other hand, the technological ethics system is taking shape. The original opposition system between nature and man will gradually evolve into a technological ethics system dominated by secular culture and enlightenment.

Stephen Hawking, a famous scientist, expressed such concerns in his book *Brief Answers to the Big Questions* that although laws can prohibit human beings from editing genes, humans cannot resist temptation." How to get rid of the shackles of capital to speed up the construction of civil society has become an urgent problem to be solved. People have great expectations for technology. As Dennis Gabor, the winner of the Nobel Prize for Physics in 1971, said that all that can be realized at the technical level deserves to be fulfilled at any moral cost. Under the joint influence of technology and markets, people begin to go astray further and further.

Many countries and business organizations have set out to establish all sorts of ethics committees to examine the acceptability of scientific and technological achievements to guarantee that the application of products meets ethical requirements. However, the root problem lies in the fact that "the essence of markets is breaking all shackles ahead with innovation", and the development of ethics always lags behind that of science and technology. What's more, human beings are becoming a creator by getting power from nature by virtue of the advancement of experimental science. However, they forget to set limits for the power. In *Humankind 2.0: Changed Bible*, American futurologist Ray Kurzweil discusses "technological singularity", which combines biotechnology, robotics and AI, and he also expresses aggressive future individualism through the study of efficiency and technology leap behavior.

In regard to the thinking of human existence, in the 1970s, Michel Foucault, a French philosopher, refuted that the humanities we understand cannot be equated to the universal propositions of humanism. In fact, it is constructed by a set of clear assumptions about "human", which are limited by history and context. Human beings, an integration of life, labor force and language, are a continuously developing "double structure of empiricism and transcendentalism".

In fact, in such an era of rapid technological development, we have to reexplore this issue from the angle of the relationships between technology, society, ethics and economy and pay more attention to social and ethical values other than economic benefits. Both values also have a decisive influence on the business itself, and this is exactly what is going on in the real business world. That is, technology and capital have changed the world, but ethics will change the trend of business and capital.

Expanding gene editing technology from the scope of "treatment" to "prevention" will blur its boundary, while "prevention" and "improvement" are just a step apart. When a family expects its offspring to have certain characteristics (high appearance, good health and high IQ), a huge crisis is looming. If ludicrous He has started a piece of history, the peril of it lies in that it is likely to bring an end to human civilization, including destroying the diversity of the human gene pool, causing eternal inequality, etc. This is also the choice and challenge that human civilization is facing. The question "What makes a human being a human being?" not only redefines the development history of our technological civilization but also defines the history of the whole "anthropocene". How to make choices will lead us to completely different evolutionary paths and civilization processes. Since it is a major historical mission of our generation, we should treat it with a particularly cautious attitude to avoid enormous risks and unpredictable future.

1.2 Man-Machine Symbiosis: Challenges and Problems in the Near Future

During the process of the development of AI in the past few years, all positive and negative problems involved are related to the same proposition: man-machine symbiosis.

It can be seen that views of the whole society on AI have been divided into two camps. One camp assumes that AI has brought new technological dividends to social development, energizing multiple industries. The possibilities of human society and the human future will eventually be reshaped through the development of AI technology, and we are becoming a semiorganic and semimechanized "cyborg". The other believes that AI technology has brought about immense ethical risks, which will impact the basic order and ethical bottom of society, leaving the person in charge of ethical dilemmas. Even in the long run, human beings may be thoroughly surpassed, backfired and replaced by AI.

In reality, unlike the mainstream deep learning methodology, the core philosophy of AI is to assume that intelligent systems can operate under constrained resource conditions. Based on this cognition, this article discusses the real dilemma between AI ethics and the philosophy of technology.

In terms of the research stage of AI ethics, the current research on AI ethics by scholars worldwide has gone through three stages. The first phase is about its necessity. Relevant research was mainly initiated in the United States. Due to American leadership in AI technology, enterprises represented by Google encountered ethical problems such as AI militarization in the early stage, which attracted the attention of industry and academia.

The second phase is the discussion of AI ethics. During this stage, both the EU and China were actively involved. For example, in April 2019, the European Commission issued a set of AI ethical principles that includes ensuring human initiative and supervision, technological robustness and security and transparency, strengthening privacy, data management and accountability, maintaining diversity, non discrimination and fairness in the application of AI systems and increasing social well-being. To date, dozens of research institutions or organizations have put forward their own ethical standards and suggestions on AI. Generally, these principles have certain universality and internal consistency and have reached a certain consensus on the establishment of ethics.

The third phase is where we are now. During this period, the studies focus on the AI ethics system and its specific connotations and application measures. Through the system of "ethical mission—ethical standards—rules for implementation", we can solve the two problems that cannot be solved at the principle level. One is "the self-execution of AI ethics", that is, how to implement the principles via cooperative operation mechanisms. The other is the "risk control of AI ethics", which plans to reduce its application risk through forward-looking deployments. In short, the AI system planning stage is the process of implementing AI ethics from theory to practice. Only in this way can we standardize the development of AI technology and constantly improve its technology evolution path.

Technologically, how does the AI ethics system adapt to the dynamic evolution of AI technology? Before answering this question, we need to determine the technical essence of AI. AI technology has different paradigms, including logical intelligence (propositional logic and first-order predicate logic), probabilistic intelligence (Bayes theorem and Bayesian network), computational intelligence (genetic algorithm and evolutionary computation), neural intelligence (machine learning and deep learning) and quantum intelligence (quantum computation and quantum machine learning).

Generally, **AI can be regarded as a technology for humans to interact with the real world through machines with the help of logical reasoning and perceptual learning on the basis of big data.** In other words, the underlying architecture that AI logic algorithms can execute is massive data. The more data resources an AI company possesses, the more competitive it will be. A professor of AI and ethics at Stanford University, Jerry Kaplan, the author of *Humans Need Not Apply: A Guide to Wealth and Work in the Age of Artificial Intelligence,* holds that a top AI company often relies on a large amount of data. Strong ones will become increasingly stronger because they can form a virtuous circle by virtue of data accumulation, iteration and automatic annotation. Machines are able to summarize laws and specific knowledge from a particular large amount of data and then apply such knowledge to real scenes to solve practical problems. That's how computer practices logical reasoning.

Since the establishment of the concept of AI in 1956, it was initially dominated by the logic school. This is primarily because logical reasoning and heuristic search avoided the deep-seated complex problems in the brain's law of thought in intelligent simulation and made major breakthroughs in key fields such as theorem proving, problem solving, pattern recognition, etc. In the early days, scientists generally believed that the essential difference between AI and traditional computer programs lies in logical reasoning. This way of thinking puts aside the microstructure of the brain and the evolution process of intelligence and only uses the process of problem solving through program or science of logic to simulate human thinking logic, so it is also classified as weak AI.

Looking back on the theory of Descartes, an important philosopher in modern theory of knowledge, real intelligence will be embodied as a "universal problem-solving ability" rather than a post synthesis of specific problem-solving ability. The fundamental feature of this universal ability is that it has the plasticity of constantly changing itself in the face of different problem contexts and has strong learning and updating abilities.

The author believes that universal problem-solving ability is a necessary and insufficient condition for machine perception. Only when the machine can not only outperform human beings on a specific problem but can it build a self-logic and learning system to give outside feedback on all kinds of universal problems can it realize machine awakening from consciousness. Correspondingly, Kant put forward the theory of mind integrating empiricism and rationalism in his book *The Critique of Pure Reason.* He divides the perceptual activities of the mind into two parts: one is perceptual ability, whose task is to pick out the original inputs of sensory information;

the other is intellectual ability, whose task is to organize those inputs into a coherent and meaningful universal experience. Kant devoted most of his energy to intellectual ability, worked out a fine model of high-order cognition, and divided intellectual ability into twelve categories with this model.

Once a machine gains intellectual ability, it obtains the ability to perceive and interact with the world, which can also be called autonomous consciousness. Autonomic awareness allows machines to deal with complex systems through self-perception and learning without preset programs. From the concept of "cognitive computing" put forward by the American psychologist and computer scientist J. C. R. Licklider, it is possible to make the computer systematically think and propose solutions to problems and realize man-machine collaboration in decision-making and control of complex situations without depending on preset programs.

We can also gradually generate the subjective impression of machine perception from a series of science fiction movies: from the German film *Metropolis* in 1927, *2001: A Space Odyssey* in 1968, to *Big Hero 6*, *Her* and *Ex Machina* in recent years. The audience generally believes that strong AI will bring us a conscious, humanoid robot whose intelligence is equal to or even better than that of human beings.

Although AI technology is not yet mature at the moment, which means we are still in the "era of weak AI", the ethical problems in AI application in real life have become quite serious. With the increasing popularity of self-driving cars, especially the occurrence of some unmanned traffic accidents, the "trolley problem" has become an issue that must be considered to ensure the safety of unmanned driving and even for the ethics of AI.

MIT launched an online test project called Moral Machine to collect and sort out public moral decision-making data. The data of 40 million moral decisions made by millions of users from 233 countries and regions reflect some global preferences: saving humans rather than animals, saving the majority at the expense of the minority, and giving priority to children's lives. However, due to the heterogeneity of geographical and cultural factors, people in different regions still have different tendencies in the choices to face the same problem. In 2018, Germany set the first ethical rule for self-driving automobiles. The rule mentions that the system must give the highest priority to human safety compared to the damage to animals or property. If the accident is unavoidable, any discrimination based on age, sex, race, physical characteristics or other differences is prohibited. For morally ambiguous events, human beings must regain control. The way of controlling the choices and behaviors of machines by implanting preset moral algorithms into machines is top-down.

In terms of solution, the designer (human) must first reach social consistency in ethical theory, analyze the information and overall program requirements necessary to implement the theory in the computer system, and then design and execute the subsystem under ethical theory. Although this top-down design approach can guarantee relative fairness on the basis of the veil of ignorance, the preset algorithms often get caught up in paradoxes in specific circumstances.

In contrast, "bottom-up" is a new thought that allows machines to derive their own moral standards through the iteration of daily rules. Machines with perceptual learning ability can summarize external information and form system behavior patterns under different situations. In the meantime, designers can encourage machines to take certain actions by establishing a reward system. Such a feedback mechanism can encourage the machine to develop ethical norms of its own in time. The process is similar to the learning experience of moral characters in childhood. By doing so, AI can truly become the subject of artificial morality and guarantee the legitimacy of its behaviors.

To explore whether machines can become moral subjects, we must think about the relationship between human beings and machines. The technology of AI has broken through the boundary between human and nonhuman entities since the enlightenment. With the growth of AI, the relationship structures between man, technology and the world have changed. Man and technology began to merge, such as the cyborg relationship and compound relationship proposed by the post phenomenological technology philosopher Verbeek. When the subject status of machine is independent of human beings, can it become a more humane responsibility subject?

Take the intelligent UAV as an example. Since the U.S. Department of Defense announced the establishment of JAIC in June 2018, and the United States has continuously accelerated the pace of AI militarization application. At the beginning of 2020, the US troops attacked and killed Qasem Soleimani, a senior Iranian general, leaving the situation in the Middle East suddenly flaring up. Media reports said the task was carried out by UAV "Reaper". With the deepening of AI technology in the field of military application, UAVs represented by "Reaper" have possessed the features of intelligence, thus triggering a new debate on the responsibility subject of war—that is, can highly intelligent UAVs better bear war responsibilities than humans as the only responsibility subject in traditional wars, leading the future war in a more humane direction?

Ronald C. Arkin of the Georgia Institute of Technology pointed out that compared with manned combat platforms, intelligent robots have the following six advantages in ensuring the justice of engagement: 1. they do not need to consider their own safety; 2. they have superhuman battlefield observation ability; 3. they are free from the influence of subjective emotions; 4. they are free from the impact of customary models; and 5. Faster information processing speed; 6. independent and objective monitoring of moral behaviors on battlefields.

Based on the above advantages, Arkin believes that the intelligent unmanned combat platform will outperform humans in the execution of humanitarian principles. However, as the current UAV autonomous system is still unstable and risky, accidents including control system failure, electronic signal interference, hacker network attack and other unforeseen circumstances on battlefields will affect its implementation of decisions in line with humanitarian regulations and even cause the killing machine on battlefields to get out of control. In addition, the dilemma of responsibility distribution caused by UAVs is also reflected in how to tackle "responsibility transfer". Backed by the high human-machine integration, the military and the government can transfer man-made responsibilities to UAVs to avoid war crimes.

The emergence of unmanned warfare will inevitably lead to profound changes in some traditional war ethics, which need to be taken seriously. Currently, some countries have proposed formulating international laws and ethical norms for military

1.2 Man-Machine Symbiosis: Challenges and Problems …

unmanned aerial systems (UASs) of increasingly higher intelligence to restrict their battlefield behaviors. On May 27, 2013, the regular meeting of the United Nations Human Rights Council also pointed out that robots are likely to accidentally kill enemies who are preparing to surrender if they are sophisticated enough to be able to judge enemy situations automatically instead of just being controlled remotely. This reminds us that we should not only use war ethics to safeguard our own interests but also change war ethics to provide legal guarantees for the application of unmanned systems or develop unmanned systems in accordance with war ethics. For example, unmanned systems should be modified to automatically identify and aim at the weapons used by the enemy, making them ineffective or destroyed, to remove the threat to their own side without killing relevant personnel. In this way, people's concerns about potential "robot killers" can be eased.

Making machines have "mind" is the pursuit of human beings, which also raises concerns about human security. This actually reflects the fear of the uncertainties of AI autonomous evolution that humans face. As Hawking said, "used as a toolkit, AI can augment our existing intelligence to open up advances in area of science and society. However, it will also bring dangers… The concern is that AI would take off on its own and redesign itself at an ever-increasing rate. Humans, who are limited by slow biological evolution, couldn't compete and would be superseded. In addition, in the future AI could develop a will of its own, a will that is in conflict with ours".

From the technical path, futurologists are worried that super AI is capable of intelligent evolution and self-replication, and then it can surpass human beings in thought, that is to say, it will achieve the technical singularity. **To understand whether AI can achieve this so-called technical singularity, we need to make it clear that the essence of intelligence contains the basic elements required by the sensory ability mentioned above**.

Let us return to the origin of the philosophy of technology. Only by deducing the future of technological development through the logic of epistemology and scientific principles of philosophy and discussing comprehensive knowledge in multiple academic fields, such as cognitive science, mental research and language modeling, can we eliminate the discussion of utopian theory or crisis theory, which are meaningless to the development of AI, and return to the path of realism to construct ethical and technical routes that are conducive to the future development of AI.

The arrival of AI makes us consider the path of human evolution. Without considering the extreme situation of human extinction in science fiction movies, there are two evolution directions for the immediate future: one is the formation of new species. A new human race can evolve on the earth or on other planets; that is, new species can be formed through directed evolution (in fact, "gene editing" is also the technical practice of such a directed evolutionary path). The other is man-machine symbiosis combining machines with the human brain to create new symbiotic species. Human beings are getting more and more "cyborg". This is an important development direction of man-machine combinations and a question that needs to be answered in the "near future" we have to face.

1.3 Data Ethics: Political Economy in the New Oil Age

Machine learning and big data are promoting the transfer of technological and commercial power throughout the world. It can be seen that companies with the highest market value in 2001 are all energy enterprises, such as GM and Exxon Mobil. However, by 2020, the most valuable companies have become those closely related to big data. It is no wonder that people say that data are oil and data are money at the moment.

The fourth industrial revolution takes big data as the core. All subsequent technological changes, such as the Internet of Things, AI and blockchain, can only be driven by big data. With the development of big data, the potential value of big data information has been continuously developed.

Data have become the "oil" of the data economy era. Countries with big data capabilities have successively implemented big data development strategies, promoted the reform of methods of production and information exchange, and expected to improve the quality of economic growth through data value. However, what cannot be ignored is that ethical issues about big data, such as data monopoly, data privacy and data information security, have also drawn great attention as the value of big data is continuously developed and verified. How to standardize and promote the use of data has become an important topic in the development of AI.

In reality, the development of information value depends on the collection of large-scale original data. The Internet, mobile communication, e-commerce, social platforms and government departments are all collecting massive data. However, what kind of personal data can be available and how to avoid data abuse are difficult to discriminate in specific practice. From the perspective of economics, this section will establish a thinking framework for readers by discussing the institutional construction and philosophical problems of data ethics.

The sharing of data and information is in an increasingly balanced stage. In the big data environment, information sharing and flow are the premise for the information value of big data to be available to the public. Without information sharing, the so-called "information island" is formed, and the value of information cannot be fully developed. At the same time, information abuse will lead to the disordered development of data, causing corresponding data ethics disputes.

Let us first look at the problems of data monopoly and privacy protection. Taking the notorious "Facebook Cambridge Analytica" as an example, Facebook was found to have manipulated American presidential election by its interest groups through the data on social media platforms, which not only sparked widespread outrage but also made us realize that we often underestimate the shaping power of big data to society. Facebook's database holds a large volume of Internet users' information, and social media giants are imperceptibly exerting influences on people's decisions in a monopolistic posture. Human beings live in a virtual and digital activity space in which they use digital technology to engage in information transmission, communication, learning, work and other activities. Each individual's words and deeds may inadvertently leave some "traces" and become an object that can be recorded and analyzed. In the era of big data, the quantified "traces" do not exist in isolation but are inextricably linked. Monopolists can produce a new set of power relations through correlation coupling and then have a profound impact on the increasingly digitalized society.

In essence, what we have learned from in the Facebook case is the "duality of information sharing": the free boundary of information sharing and the value expansion of information islands. This contradiction is almost an endogenous problem of data value. Facebook provides nearly unlimited information sharing for all users, but it also brings problems such as privacy and data monopoly.

From the perspective of the logic of digital economics, the privacy and monopoly of big data is the constraint mechanism of science and technology ethics in the era of information sharing. The basic logic behind it is the boundary of information sharing and the fair distribution of information value. We should think about how to establish a constraint mechanism for data ethics to avoid corresponding risks while ensuring that the value of big data information is tapped.

Next, before we discuss the formulation of data protection rules according to the relationship between system and ethics, we need to look at Google's privacy infringement case. In June 2017, Google blocked its competitors' shopping comparison websites for its own promotion in search results. The move violated Article 102 of the *Treaty on the Functioning of the European Union (TFEU)* for the abuse of its market dominant position, and Google was fined a huge amount by the European Commission. In this case, relevant institutions creatively put forward the concept of "the right to be forgotten", which is used to indicate that people have the right to require relevant service providers to delete personal information such as data traces left on the Internet in the era of the digital economy.

From the viewpoint of privacy protection, the case of Google data privacy provides a legal framework for data protection in three aspects: anyone has the basic rights and freedom to access and process personal data, the controller of personal data must bear the legal obligations and responsibilities of personal information, and all EU members must establish special data protection institutions. In fact, it can be seen that the verdict returned by the European Union Court of Justice over the case of *Google Spain SL and Google Inc. v AEPD* not only achieved the above three objectives but also proposed relevant opinions on the innovation of adopting extensive interpretation in the provisions on the protection of personal data privacy in the *EU Data Protection Directive*. The *General Data Protection Regulation (GDPR)* launched by the EU after 2018 is obviously affected by the relevant cases, prompting people to focus on the freedom of information sharing and ethical limits of data space as public fields.

Facing the challenges related to data ethics, governments have formulated many laws and regulations relating to privacy and data security. The European Union launched the *Directive on Privacy and Electronic Communications, Directive 2002/58/EC*) in 2002. The United States protected people's right to privacy through the relevant precedents of *The Fourth Amendments to the Constitution of the United States* and *The Fourteenth Amendment to the United States Constitution*. The U.S. also issued the *Consumer Data Privacy in a Networked World* in 2012, the *Protecting the Privacy of Customers of Broadband and other Telecommunications Services* in

2016 and some other regulations. Moreover, many international cooperation organizations are also carrying out relevant institutional construction, such as the European Federation of Data Protection Organizations (CEPDO), the International Association of Privacy Professionals (IAPP) and other institutions.

It is worth mentioning that the *Research Report on Data and Competition Policies* issued by the Japan Fair Trade Commission (JFTC) in June 2017 organized and introduced many issues related to data, such as the changes in the use environment and status of data, the impact assessment methods of mobile phones and data use on competition, data collection and use behavior. The document focuses on a large number of frontier issues of the data system (such as privacy considerations in the framework of competition law, data material blockade, etc.). Moreover, Japan's report highlights the aspects of data collection and use and divides data into personal data, industrial data and public data. In particular, the research on the latter two types of data was very cutting-edge and detailed.

For personal data, the report proposes that the concepts of data and information are convergent (the cross perspective of data ethics and information ethics) and mainly discusses relevant behaviors in the social network market. In terms of industrial data, the report emphasizes the concept of data hoarding and points out that monopoly or oligopoly enterprises may restrict data access or data collecting channels to achieve data hoarding. Regarding public data, the research focuses on how to maximize the value application of government institutions and public data. By this means, the report analyzes the necessity of defining the data trading markets and puts forward crucial perspectives on the M&A and a review of the data.

Although these directives or regulations escort data ethics to a certain extent, from the angle of institutional economics, the crises of data supervision still exist. One of the reasons is that the modern anti-monopoly law based on the theory of Chicago School mainly focuses on three types of behaviors: mere horizontal fixed prices and market segmentation monopoly agreements, bilateral monopoly and horizontal merger of monopoly, and limited exclusive behaviors. Being affected by the above phenomenon, regulators tend to ignore the maintenance of data competition in cross industry mergers, and multiple data-driven mergers and acquisitions have not been included in the concentrated reviews of operators, fully reflecting the crisis of dealing with problems relevant to data monopoly on the multisided markets under the mindset of modern anti-monopoly law.

Moreover, the antitrust investigation on the abuse of market dominant position by platforms turns out to be a kind of lagging and passive post review. To promote the long-term and healthy development of the data industry, regulators should not simply ask enterprises to accept monetary compensation or provide data beyond the necessary scope on the grounds of breaking the enterprise's "data monopoly". Therefore, in addition to strengthening the prior review of cross-industry data M&As and integration on platforms, optimizing the data sharing mechanisms is also the key to promoting the development of big data.

In the field of data ethics, the traditional economic mode mostly represented by physical resources takes land, natural resources, population and capital as production factors, while key information and value elements in the digital economy are generally reflected in the production, storage, circulation and application of data resources. The form of virtual resources not only expands the application width and depth of factor resources but also combines with the traditional production factors in the real economy to form new economic paradigms such as AI, robots, blockchain, digital finance and so on, further energizing traditional products and services of the real economy in quality, efficiency and benefits to double the benefits of the real economy. Data resources have become the most important production factor in the development of the digital economy, but the data come from every individual participating in production activities in the real world. Under the infiltration of various technologies and forces, some ethical problems have been completely unmasked, mainly in the following aspects:

- Information security issues: the data industry chain is interlocking and intricate, and the risk of uncertainty in the acquisition terminal, processing node, storage medium and transmission path of data analyses is high.
- Personal rights and interests issues: The abuse of data leads to privacy overdraft, and personal dignity is more likely to be degraded.
- Business ethics issues: AI products may induce users to actively overdraw personal data, and human beings may lose control over the boundaries of privacy and sensitive data.
- The principle of informed consent is becoming dispensable in the context of ubiquitous sensors and potential data development value.
- Intellectual property issues: big data is aggravating the contradiction between intellectual property and free data sharing on the Internet.

We hope that members of society can equally obtain data resources according to their needs and acquire corresponding wealth distribution based on their contributions to production activities. However, in reality, due to the contradiction between data-driven productivity and productive relations, individual users and platform enterprises often oppose each other in the application of data resource distribution, and there also exist contradictions such as data monopoly among enterprises. Specifically, it can be deconstructed into the following four aspects:

First, "who owns the data?"—the ownership of data property has always been the focus of debates in the industry, especially for those data transactions that remove the attributes of personal identities, as well as the data produced by individuals who are collected and stored after desensitization treatment by enterprises or government departments. There is still no consensus on whether data ownership should belong to individuals, enterprises or the government.

Second, "who is using the data?"—government and enterprise are the two principal users of large-scale data in the current digital economy era. The government collects a large amount of data through various systems, including public service websites, digital government platforms and the "Integrated Online Platform", while enterprises gather users' information by providing services for users and obtaining multidimensional trends and regular features through data analyses to precisely improve and optimize their services and user experiences. For this reason, the convenience, nonrivalness and low-cost replication of data storage and transmission reflected in the above application process also make it difficult to protect data property rights. Data can never get rid of being occupied, stolen and abused illegally even with clear property rights. Moreover, as technologies diffuse, people come to realize the value characteristics of data. Data theft technologies such as web crawlers and credential stuffing attacks are rampant. Whether public privacy, government governance or national security is more vulnerable to the threats of improper use, such as privacy and data theft and abuse, than ever before, seriously infringing on the property rights of data owners and causing damage to the scarcity of data.

Third, "How much data is available?"—as the source of wealth and value in the digital economy, the data generated by individual consumers are not only the basic source of profits and value of platform enterprises but also the key element for digital public services and digital governance of government. However, currently, the unclear ownership of personal data leads to extreme situations such as abuse or excessive restriction on the use of personal data, which place the public individual interests and the interests of platforms or public organizations in a contradiction of binary opposition.

Fourth, "who should get the data revenue?"—the considerable economic benefits brought by using data to optimize products and services. The distribution between data producers (individuals), collectors and processors (enterprises and government) affects the interests of many subjects. Although the current judicial decisions prefer to distribute the data revenue to the collectors, producer and actual controllers at the link of secondary exploitation of data—enterprises. However, in some public services, especially data scenarios relating to government affairs, can the government as a data user and individual data producers have the right to enjoy legal benefits of data without the support of judicial decisions?

The above is the discussions on science and technology ethics of big data from the points of information ethics, institutional assumptions and philosophy. Whether it be big data, AI or gene editing technology, they are essentially subversive technologies in the digintelligence era, such as "every corn has two sides". When we look at the ethical issues of science and technology, we should not only pay attention to their technological advantages but also examine their disadvantages through other angles, such as economics and philosophy. Only by doing so can we have the chance to understand their application paradigms in practice as well as their impact on the basic logic of social operation.

Only by this means can we continue our accumulation of positive value in shaping modernity and apply more and better science and technology to the expansion and development of human civilization after entering the "digintelligence risk society". In the meantime, the risks brought by technical ethics, the conflict between humans and nature, the continuation of the human race and the spiritual feelings of loneliness and wandering caused by modernity should all be controlled at an acceptable level. It is also an important and essential problem that the author believes the digital economy era is facing.

Chapter 2
Ethical Enlightenment in the Age of Intelligence

If the Enlightenment Movement can be seen as the impetus for rational reflection and value reconstruction in specific fields, the trend of thought "tech for social good" driven by China's science and technology companies in the past few years can be regarded as an obvious mark of China's "ethical enlightenment" movement in the smart era.

The proposal of the idea of "tech for social good" in this age primarily stems from the ethical challenges we have encountered when promoting the development of the digital economy, such as the repeated big data discrimination on travel websites and the security risks brought by car sharing. We have seen that these technical ethical issues are challenging people's moral bottom line and their ethical concepts of digital life. How to delimit the red line when technology promotes business development and how to understand the role that the concept of "tech for social good" plays in creating a good social ethics orientation and soft landing mechanism for scientific and technological innovation are the propositions we need to think about.

Given that AI technology is one of the important technologies to solve human poverty and inequality and may create and expand the digital divide in its development, it is very necessary for us to understand technical ethics from the angle of economics—we should fundamentally consider the challenge of ethics to the underlying logic of social development rather than just the impacts of technology on economy and society. Technology enterprises take tech for social good as their mission, which actually determines the new value dimensions of assessment techniques, products and services. The change also means that technical innovation needs to consider not only commercial value but also other aspects, such as ethics and social responsibility. Obviously, the move is not only an enlightenment but also a cognitive upgrade.

The core of AI ethics research is to "put people first". We should make AI work for human well-being by coordinating human ethical values with those of AI. Furthermore, human beings should first reach some "consensus" on their own ethical values before discussing the development framework and basic governance principles of AI ethics.

© The Author(s), under exclusive license to Springer Nature Singapore Pte Ltd. 2022
Z. Liu and Y. Zheng, *AI Ethics and Governance*,
https://doi.org/10.1007/978-981-19-2531-3_2

Therefore, the focus of this chapter is to discuss these already formed "human ethical consensus" and their corresponding origins and understand these ideas on the basis of modernity. As Kant pointed out, "Nothing is required for public enlightenment, however, except freedom; the freedom to use reason publicly in all matters".

The first part of this book aims to tell readers that the **ethical problems of AI (including all technical ethical problems) are actually human problems brought by "modernity" and "modernization"**.

It is worth mentioning the "impoverishment" of modern economic theory criticized by Amartya Sen, the Nobel Prize winner of the economy. On the one hand, mainstream neoclassical economics today is analyzed and constructed under an ideal theoretical assumption of zero transaction fees, so it differs greatly from the real economic world of human life. On the other hand, although the new institutional economics theory has introduced variables such as "transaction fee" to explore the institutional arrangement of market operation, "ethics" of systems has never become a consideration, and the moral basis and ethical thinking seem to have never been incorporated into the research framework of economics. Therefore, this is also the ethical difficulty that we face in the digital economy—since the traditional economic perspective has never taken ethics into consideration, how do we construct ethics in the era of the digital economy and AI?

If we do not look at the paths and roles of institutional changes from the perspective of ethics, we cannot understand the reasons for the existence of the "non-Pareto effect" or "non-Nash-Sutcliffe efficiency" in economic operation. It is also impossible for us to understand why most people follow certain ethical norms to coordinate and cooperate with each other in the community.

The real economic world is Hayek's "extended order of human cooperation", rather than the "a war of all against all" discussed by Hobbes in *Leviathan*. In his paper "Conjecture on the Beginning of Human History", Kant mentioned his views on history, human nature and human society, which may help us correctly understand the problems we are facing. He made a guess at the initial development history of human freedom and gave us a poetic interpretation. At first, human beings only had animal instincts. He had to rely on the guidance of instincts and put himself at the mercy of God. However, gradually driven by reason, he expanded his range of food. Then, the reason soon made him notice its existence. Human beings began to seek to transcend the boundaries of instincts and expand their knowledge. This marks that reason began to control the consciousness of impulse. After that, mankind began to learn to predict the future, that is, how to prepare for long-term goals, which is the decisive factor of human strength. Finally, human beings realized that human is the real purpose of nature, so they take whatever they can as tools and means with the exception of human beings themselves. Thus, they found a way to change the evolution of nature. To put it another way, historically, behind the human evil lies the purpose of good". Therefore, when we look at the ethical problems in the intelligent era, we also need to adhere to this human-centered principle. To realize self enlightenment, we need to have the courage to use our own rationality to confront unknown problems.

2.1 Development of Human Consciousness and Machine Consciousness

Before discussing the ethics of AI, we must first discuss the consciousness of AI from a philosophical perspective. Yuval Noah Harari, the author of *Sapiens: A Brief History of Humankind*, once said, "intelligence and consciousness are very different things. Intelligence is the ability to solve problems, while consciousness is the ability to feel things such as pain, joy, love, and anger". Obviously, ethics is a proposition related to consciousness, so whether there is an essential difference between machines and humans determines the significance of ethical problems.

In this chapter, we discuss the philosophical significance of human consciousness and machine consciousness from the perspective of Eastern and Western philosophy and the necessity of ethical issues from the uniqueness of human nature. Finally, we try to understand the development of AI from generalized consciousness and look at the impacts of such a technological paradigm on human civilization.

First, let us discuss the relationship between intelligence and consciousness. We can compare the differences between them from philosophical concepts. "Intelligence" tends to treat and deal with problems in a purposeful, objective and effective way in hopes of obtaining answers to questions. "Consciousness" refers to the subject's ability to feel pain and emotions. It is the ability of an individual to construct meaning after being touched physically and mentally. Therefore, consciousness represents a certain meaning that can be embodied in certain objects. For example, the phenomenon of the "phantom limb" in medicine means that individual consciousness can be given meaning subjectively whether there is an object or not.

It can be seen that philosophers basically follow two paths when discussing human nature: "intelligence" and "consciousness". It is precisely because different civilizations and societies have different understandings of intelligence and consciousness that the East and the West have different understanding directions of contemporary technological philosophy, showing the evolution paths of philosophy of AI from different perspectives.

By comparing the relevant contents of Chinese and Western philosophy, we find that the West pays more attention to the philosophical significance of "intelligence". People answer through intelligence when discussing the proposition of "why human beings are human beings". Therefore, the main goal of "robot ethics" is to create a robot that can act in accordance with ideal moral principles and ethical laws. Thus, ethical dimensions need to be added to robot intelligence.

(1) In the design of robots, ethical norms and moral laws can be embedded into robots according to human conceptions so that they can carry out corresponding behaviors according to the moral ways set by engineers.
(2) The ideal moral principles of the robot should be set up, including the corresponding moral dilemma, cases of correct feedback and the algorithm of abstract ethical principles for each approval project, so that the robot can guide its behaviors according to the corresponding examples.

In contrast with western "intelligence", oriental philosophy mainly discusses ethics and human nature from the level of consciousness, and one typical example is Chinese Confucian moral philosophy.

The ethical representatives of Confucianism include "Mencius" and "Hsun Tzu". Mencius believes that what makes humans unique is that all men have a mind that cannot bear to see the sufferings of others. Cultivation of kindheartedness can make humans have real human nature, promoting human society to a moral life. Hsun Tzu believes that people become moral by acquiring social norms; that is, they can act morally through continuous social practice, which means they can obtain ethics through social procedures.

Therefore, both Mencius and Hsun Tzu emphasize that human moral consciousness is crucial. From this angle, Hsun Tzu's idea is more suited to the development of contemporary AI. He believes that the keys to human consciousness are "righteousness" and "group". The former defines the appropriate consciousness of morality of the human theme, and the latter defines the human ability to form groups, and both emphasize the importance of ethical rules in the social system.

We can see that compared with Western philosophy, Confucian philosophy can better deal with ethical challenges in the field of AI. For example, Kant's ethics takes rationality as the premise, and he believes everyone possesses rationality and can practice it, while Confucian school discusses how different people promote ethical practice according to their actual situations under the assumption that everyone can be a gentleman. In other words, Confucianism leaves more room to embrace AI as a moral subject, and AI needs not to be treated as fully equivalent as humans.

Then, let us discuss the similarities and differences between AI and human consciousness. One thing worth mentioning here is Moravec's paradox. It was proposed by Hans Moravec, Rodney Brooks, Marvin Minsky, et al. in the 1980s. Moravec's paradox points out that, contrary to traditional assumptions, for computers, high-level reasoning requires very little computation, but low-level sensorimotor skills require enormous computational resources.

Scientists have found that it is comparatively easy to make computers exhibit adult-level performance on intelligence tests or playing checkers, and it is difficult or impossible to give them the skills of a one-year-old in regard to perception and mobility. Because a large number of perceived behavioral abilities depend on considerable tacit knowledge, human emotion and certain self-consciousness, they are difficult to algorithmize. To put it another way, the AI we face belongs to the "unconscious machine", and its characters can be reflected in three aspects: (1) AI is not able to establish a connection with the external world since it is not capable of understanding the meanings of symbols and the physical world, especially the meanings of human languages; (2) AI has no consciousness, an important feature of human beings as a unique life form, and related experiences; and (3) AI lacks autonomy and self-awareness. What we see is the coupling of human-computer interaction.

At present, research on human and AI consciousness has become an important topic. Scientists are exploring how to help AI gain consciousness. One is to construct

2.1 Development of Human Consciousness and Machine Consciousness

consciousness through algorithms such as symbolic calculation and statistical calculation; the other is through brain-like research, that is, we can carry out relevant research under the inspiration of the study of human brain structure and working mechanisms.

The most well-known is the Human Brain Project (HBP) launched by the European Union. In 2013, the EU led the ten-year HBP, with 135 cooperation institutions from 26 countries becoming involved. They set an extremely high goal for this project with a high-risk investment of up to 1.3 billion euros. They planned to develop an information and communication technology platform that is committed to neuroinformatics, brain simulation, high-performance computing, medical informatics, neural morphology computing and neurorobot research. The research aimed to realize AI by simulating brain functions with these supercomputer technologies.

The project can be divided into three stages. The first stage is two-and-a-half years of the "climbing stage", during which they established a preliminary version of the ICT (information and communication technology) platform and implanted it with some strategically selected data to prepare a platform for researchers inside or outside the project. The second stage is the "operation stage" for the next four to four and a half years. The targets of this period of time are to produce more strategic data and add more functions by strengthening platform application and exhibit the value of the platform to basic neuroscience research, drug development and application as well as future computing technology. The last three-year "sustainable stage" is to ensure that the project is financially independent and sustainable and can become a permanent asset of European science and industry.

Of course, this plan seems to have failed thus far. In contrast, China's brain science research turned out to be more practical and reliable. In 2016, China joined the global competition of the "brain project". The China Brain Project has two orientations: brain science research guided by the exploration of brain secrets and brain diseases and brain-like research directed by the establishment and development of AI technology. The China Brain Project took the study of the neural principle of brain cognition as the "main body" and the development of new approaches in the diagnosis and treatment of major brain diseases and new brain-machine intelligence technologies as its "two wings". It aims to achieve international advanced results in the three frontiers of brain science, including brain science, early diagnosis and intervention of brain diseases and brain-like intelligent devices, in the next 15 years. The ongoing scientific research plan based on China's actual situation and international experience has been recognized worldwide.

We will summarize the research and development of human and machine consciousness here. Minsky once said "man is just an emotional machine in the flesh". Scientists proposed that the construction of conscious machines needs to meet the following conditions: perceptual state, conscious ability, attention ability, planning ability and emotion. Machines must have emotions to be conscious. That is, they know how to make value or moral judgments and behave morally. The AI ethics we are discussing need to learn human ethics and value judgments from bottom to top to generate spontaneous moral emotion and moral awareness. It is far from enough to merely implant human norms into machine systems from top to bottom to acquire the ability to make ethical decisions that will satisfy actual needs and do no harm to humans.

Building a good ecological and social environment for human-computer interaction is a specific challenge we are facing. We should be particularly vigilant to avoid machines with malicious consciousness. If we can produce AI products with autonomy and self-consciousness in one day, the subject rights of the machine will also become a problem. We need to consider the future development of humans and machines, especially the construction of related morality and ethics, from the level of individual subjects, the individual self and the social technology system.

2.2 Looking at AI from the Differences Between the East and the West

Let us return to the process of human civilization after the discussions on AI and human consciousness. As the autonomy of AI continues to increase, its decisive role in human production activities is becoming gradually prominent. However, under the impacts of AI, traditional ethical concepts and moral theories have difficulties in applicability, and some propositions may even fall into the vacuum of moral philosophy, thereby triggering disputes.

Meanwhile, such risks also reversely force ethical and philosophical researchers to think more deeply. We should revise and optimize the current moral concepts and ethical theories based on the experience of sages and the cultural contexts and cognitive characteristics from different perspectives of East and West to put forward a philosophical moral system that can better adapt to the age of global intelligence. From multiple perspectives, we can explore the speculations of moral philosophy about AI at various times vertically and explore the ethical development of AI horizontally in the different cultural contexts of East and West, exhibiting the extension and expansion of traditional moral philosophy.

In essence, AI is the product of human beings and human civilization. Therefore, let us first think about the question of **whether there are similarities and differences in the thinking perspectives of Eastern and Western civilizations on AI and whether we can reflect on the development of AI from the angle of civilization**. Generally, to understand the challenges of AI, we can first think about the process of human civilization before taking a further look at human nature, human values, human evolution and alienation.

Let us start from Western civilization. Generally, the philosophical research on AI in the context of Western civilization came from the article "*Computing Machinery and Intelligence*" published by Alan Turing in 1950. Since then, Western philosophers and computer scientists have begun to be keen on discussing AI. They expect that machines can become intelligent as humans. They have human-like thoughts, feelings, instinctive reactions and creativity. From the origin of more ancient civilization, the ancient Greek tradition defined human beings as creatures with rational thinking. Therefore, how to develop rational intelligence is the main path for major algorithms in expanding intelligence.

How to develop the "rational intelligence" of machines, that is, how to make machines have ethical attributes, has become the focus of applied ethics research. Multiple paths have emerged surrounding the topic, including those around value-sensitive design or those based on regional cultures. Among them, "humanism" centered on human rights and "technicism" are the most typical. The latter mainly focus on how to achieve the targets that contribute to human moral goals, including safety, credibility, reliability and controllability. through technology.

"Anthropocentrism" holds that man is not a part of nature but a superior existence. Homo sapiens has unique rationality, self-consciousness and subjective personality and occupies a higher position in the world than any other animals, plants or even any life being. After human society entered the era of industrialization and globalization, anthropocentrism developed to its peak. We should put human dignity and well-being at the core when considering AI ethics from the perspective of "anthropocentrism". However, the inherent humanistic presupposition of this path leads to discussions on AI ethics falling into a metaphysical dilemma. For example, when ethical principles are embedded in artificial intelligence agents, difficulties such as the protection mechanism of human rights, the challenges of interpretation and the plight of beyond the minimum standard, etc., will occur.

In this field, a representative work is the "*Robot Series*" written by Isaac Asimov, a master of modern American science fiction, and the "Three Laws of Robotics" presented in this work, which are considered by many scholars as the theoretical basis of robot ethics and important principles for the construction of robot ethics. Isaac Asimov put forward it in "*I, Robot*" in 1950, aiming to ethically regulate future AI. "Three Laws of Robotics" are as follows:

- The first law: a robot may not injure a human being, or through inaction, allow a human being to come to harm.
- The second law: a robot must obey orders given it by human beings except where such orders would conflict with the first law.
- The third law: a robot must protect its own existence as long as such protection does not conflict with the first or second law.

What is the essence of AI different from previous technologies? In short, it is more crucial than tools. Previous technologies are just service tools for human beings, but AI is "intelligence" instead of being just a tool. It can be said that the functional understanding of AI is an important method, but one limitation of this understanding is that it sparks many people's worry and fear about the impact of functional evolution—the proposal of "Three Laws of Robotics" represents people's irrational fear of AI such as robots with emotional and empathic capabilities. Although robots can abide by the set of laws, they can live longer than humans, and their performance, tolerance and firmness under some extreme environments are better than those of humans. All these strengths may provoke human jealousy and fear and make people want to destroy them.

In the West, technicism is closely connected with AI technology and is mostly valued by technical experts. Targeting the robustness and nonparaphrase of second-generation AI, technicism discusses how to realize moral purposes conducive to human beings through technology, such as safety, trustworthiness, reliability and controllability. Two typical representatives of this aspect are reliable AI and interpretable AI.

Moreover, we can also temporarily put aside the current mainstream Westernized perspective and understand AI from the perspective of oriental civilization, which is a meaningful ideological experiment.

Corresponding to the thinking on the "human nature" of AI from the perspective of Western humanism and technicism, the Confucianism represented by Confucius in Chinese civilization defines man as a social being in interpersonal relations and believes that man can recognize each other as human beings (the definition of "humanity") and reflect on his own behaviors, values and thoughts (the way of "introspection") are important embodiments of being a social person. From a philosophical viewpoint, the emergence of AI means that human beings are creating new species. If the Enlightenment Movement marks the departure of human beings from nature to become independent beings, then the existence of AI can be regarded as a significant practice of the movement. To put it another way, after people obtained the dominant power in the sense of ontology, they began to change the process and significance of civilization.

Some Western scholars believe that cutting-edge scientific and technological progress such as AI may pose a threat to human survival, while unlike the West's excessive panic over AI, the Chinese take more inclusive and receptive attitudes toward it. As such, an interesting phenomenon, we can start with the three characteristics of ancient Chinese philosophy to help us understand why the Chinese are calmer in the face of the development of AI technology.

We note that Chinese Confucianism and Taoism were the main philosophical concepts in ancient times. Taoism holds that human nature is temporary. If a man is not obsessed with any object, then he can unify with all things on earth and realize the nature of things rather than regarding them as objective materialized objects. Confucianism believes that the family ethical relationship is the starting point of all values. The thought that "benevolence means to love others, and the greatest love for people is the love for one's parents" means the formation of human nature is to gradually spread love from those closest to you to others, even to the world or everything. It can be said that in the world outlook of oriental philosophy, the concept of non-anthropocentrism—the "trinity of heaven, earth and man"—is a classic framework for Chinese people to understand the relationship between man, nature and society.

The concept originally came from *The book of Changes* in ancient China. According to the theory in the book, the three elements, nature, earth and man, as well as the two forces of yin and yang contained in them, are the most basic elements of the universe, in which nature evolves, human beings develop and society progresses. On the grounds of this theoretical framework, human beings are essentially a part of nature and are closely linked with nature. Only by following the laws of nature and respecting the unity of man and nature can human beings multiply and thrive.

As individuals living on the earth, human beings have the unique ability to learn from nature. By virtue of the power of all things in nature, human beings make the world in which they live vibrant and sustainable and then comprehend the "Tao" and carry forward it in the world; that is, "man can make use of natural conditions to nurture all things on the planet through his intelligence". From the perspective of Confucianism, "Tao" means that people should follow the moral teachings of benevolence, righteousness and integrity. Although Confucianism encourages people to enter into society positively, it also advocates that people should respect and even advocate the laws of nature and follow the laws of nature when dealing with specific affairs rather than exploiting nature intemperately. Human beings should understand the information of heaven and earth from various natural phenomena, such as seasonal changes, and act in accordance with the "Tao".

In addition, Confucianism advocates that people should adjust their thoughts and behaviors according to the times and places. Mencius once praised Confucius as "the sage of his time", which means that Confucius is a sage who can deal with the changes of the times. Therefore, Confucianism is not a rigid dogma but a set of wisdom that can be applied to different times and situations. With extremely rich understandings of human beings, Confucian tradition emphasizes the natural life order formed by interpersonal relations in society and extends such emotion to the generalized "benevolence" of all things in the world.

Correspondingly, the concept of "harmony between man and nature" has gained a more prominent position in Taoist ideas. Lao Tzu, the founder of Taoism, points out in *Tao Te Ching* that "I am abstracted from the world, the world from nature, nature from the way, and the way from what is beneath abstraction". Tao is embodied in the three forces of heaven, earth and man. The mutual integration and supplementation of natural law, universal law and humanity can make heaven, earth and man coexist in harmony. Chuang Tzu, a Chinese philosopher who lived in the 4th century BC, further developed the thought of "harmony between man and nature". He believes that heaven, earth and man came into being at the same time, and the existence of the universe and man is one.

In conclusion, we can assume that ancient Chinese philosophy did not give man a unique paramount status in the universe. As a developing advanced technology, AI is not the product of nature. From the perspective of "harmony between man and nature", the development of AI should be limited to a reasonable range and guided and constrained in an orderly way to realize respect for the natural attributes of life. It is precisely because they are deeply influenced by their own nonanthropocentric philosophical tradition that Chinese people do not feel that their survival is being threatened by the development of AI technology; thus, they are not as fearful as Westerners.

On the one hand, many Chinese thinkers are unconvinced that artificial intelligence can surpass human intelligence one day; on the other hand, it is not uncommon for machines or animals to outperform humans in some aspects. After the inheritance of super life legends such as immortals in Taoist culture, AI or some digital forms of intelligence can be considered a super life form in the eyes of Chinese. Some Confucians and Taoist scholars hope that AI can be integrated into the order constructed by human beings in the future. They have begun to take AI as a partner or a friend of human beings.

In addition to nonanthropocentrism, Chinese culture often holds a relatively open mind in the face of uncertainties and changes. Compared with the strong panics that arise from the development of AI technology in the West, China shows a higher tolerance for uncertainties and changes due to the influence of *The Book of Changes* (*Zhou Yi*). According to the key point of the classic, "a constant state of change" is the basic form of the existence of the universe, while "existence" with static characteristics is not the essence of cosmic existence. This idea of static existence was widely recognized in the European ideological community in the 20th century. Influenced by *The Book of Changes*, Confucianism believes that people should actively predict and deal with changes, which reflects the "humanistic vitality" of Confucianism. German sinologist Richard Wilhelm (1873–1930) preached in China in the late 19th and early 20th centuries. He was the first person to translate the book into a Western language. "There is no so-called eternal plight in this world; everything is changing. Therefore, even if we are facing an extremely difficult situation, we should do something to promote the formation of a new situation in the future" he once said.

Aside from Confucianism, since the Han Dynasty, Taoism has been characterized by its willingness to keep pace with the times and take the initiative to reform according to actual situations. Chuang Tzu's viewpoint of "keeping pace with the times and rejecting stubbornness and pedantry" has gained an important status in contemporary Chinese culture. From the perspective of ancient Chinese philosophy, uncertainties and changes are not problems to be solved but an important part of the laws of nature and a constant in the universe. In Buddhist thought, impermanence is a very vital concept, and the essence of the reality we live in is illusory. The explanation makes many changes in the world more insignificant. The way of thinking in Buddhism also makes the Chinese take a more tolerant attitude toward AI.

Finally, the introspection, self-cultivation and consciousness advocated in Chinese philosophy also drive the Chinese to look back on their past when faced with AI and realize that perhaps the crux of the problem lies in ourselves. The common feature of the three ideological schools, Confucianism, Taoism and Buddhism, is that people should be self-disciplined, constantly reflect on themselves, and make endless efforts to achieve inner holiness. Moreover, these schools of thought assume that everyone should start with his own self-cultivation to build a good society, and a good society is hard to realize without good self-cultivation.

Therefore, in the eyes of Chinese philosophers, we should examine ourselves and obtain inspiration from the evolutionary history of human society when discussing the future direction of human technological development and debating the survival threat brought by technological progress. In other words, we need to review our past and realize that the key to the problem may lie in ourselves. We can never produce morally satisfactory AI products unless we can reflect on our morality and assume our responsibilities.

As humans are facing an increasing number of global challenges, we may broaden our thinking and obtain inspiration from ancient Chinese philosophical traditions. It's time to get rid of the zero-sum game mentality, the preference for maximizing personal wealth, and fully unrestricted individualism. To create "human friendly" artificial intelligence or other cutting-edge technology products, we must work to build a global system that is inclusive, harmonious and compassionate to mankind. We can see that human self-cognition is constantly enriched through practical activities, and technological progress is expanding people's abilities and the "cognitive scale" of themselves. Therefore, it is necessary to conduct a more profound study on the philosophical values and connotations of the east in this process to obtain a new understanding of the ethical relationship between man and machine in an intelligent society, which is also the core of the discussion in this section.

2.3 Ethics and Moral Responsibilities of AI

As discussed above, we have benefited greatly from the development of AI, but we should also attach importance to the many potential ethical, moral and legal risks. For example, there are still no clear answers to a series of questions, such as the definition of medical accident responsibility of "Leonardo Da Vinci" surgical robot, the ownership of copyright of "XiaoIce' poetry collection, and the ownership and responsibility of human death caused by Uber unmanned motor vehicles. To avoid the moral hazards of AI, the idea of a moral machine has begun to emerge and received much support, but the view is based on emotion rather than rationality. The proposal of machine morality stems from human concern and panic about the development of AI technology. We hope that machines can be so moral that they will not harm us while serving us.

Machine morality is the expansion of human morality in the era of AI, while artificial intelligence is artifactitious. Whether AI has human sociality is still controversial in the academic community. Additionally, the study of AI at the legal level cannot be separated from the basic category of how AI bears responsibility. The premise to solve this problem is to clarify the legal status of AI and its relationships with humans, especially the impacts and challenges of future strong AI with self-awareness and independent decision-making ability on the existing judicial system, the right of personality, the subject status and legitimate rights of humans.

Compared with the practice of traditional ethics—the moral practice of "man", the moral practice of intelligent machines such as autonomous driving vehicles has significant particularities. First, machine behaviors are determined by the computational process of artificial design instead of a natural causal process. This distinguishes machine behaviors from ordinary accidents or natural disasters. Second, machine behaviors are not determined by its users to a great extent, which makes the machine different from the tools in the traditional sense. Finally, for machines with wide applicability and the ability to deal with complex situations, especially those with learning ability, it is difficult or impossible for designers and manufacturers to accurately predict or control their behaviors.

Thus, traditional responsibility partition attribution is not suitable for intelligent machines due to these particularities. Take a self-driving automobile as an example. If a highly reliable autonomous vehicle accidentally violates traffic rules or causes traffic accidents, then it is not convincing to regard the incident as an accident, requiring passengers or designers to bear the corresponding responsibilities. To endow machines with moral subject status is to make them responsible for their own behaviors and decisions. In the long run, machines are able to be responsible for their behaviors and decisions only if they have moral subjectivity, which means machines have free will and can behave freely. It is reasonable to attribute moral responsibility to machines only when machines can act freely.

AI has the following characteristics that are highly similar to those of human beings: 1. Rational conditions—AI can independently complete the set tasks based on the algorithm model obtained from training and learning. At the same time, it can also realize unsupervised learning and evolution according to the real-time feedback data, which is similar to human learning behavior. 2. Moral internalization—human moral code of conduct is not something we are born with but something acquires in the environment. If the code of ethics can be quantified into data that can be recognized by the computer and input into AI, it can also be acquired based on its own learning characteristics. 3. Artificial intelligence has the possibility of enjoying corresponding rights and bearing related obligations and responsibilities. 4. Strong AI may have "free will" similar to human beings and generate human-like production behaviors through the mode of "cognition-learning-creation".

The reason people have personality is that they have rationality. AI has a human-like rational essence. In the eyes of scholars who oppose "Anthropocentrism", morally, we should treat all rational beings like human beings. As long as this rational relationship exists, no matter how different it is from Homo sapiens in biology, it is wrong to disrespect its human nature or treat it in an inhumane way. If we fail to follow this principle, on what grounds can we treat all intelligent beings equally? The premise of establishing such a rational relationship is that the object can obey morality like human beings and cause no harm to society, or the harm is generally within control. Only by doing so can we judge that the AI product has moral ability and meets the standards established based on ethical personalism. However, those unmoral AI products that may threaten the interests of the human race should not be regarded as moral subjects.

It should be noted that whether AI can be considered a moral subject is not necessarily related to the correct values that AI should follow. Even if AI does not have the status of moral subject, it has to observe correct ethics and values. Second, with the continuous expansion of the ability of AI, it bears greater responsibility. Meanwhile, restrictions on moral subjects will also limit responsibilities that humans should not bear. Finally, before judging that AI has moral ability, reliable standards should be set up to assist us in judgment.

2.3 Ethics and Moral Responsibilities of AI

The necessary and insufficient conditions for AI with moral subject status are cognitive ability, intent capacity and moral ability. Cognitive ability mainly refers to the human-like ability of perceiving the objective material world and spiritual world. It covers a wide range of elements, including memory, curiosity, association, perception, introspection, imagination, intention, and self-consciousness. Cognitive ability can be embodied as a process by which AI receives information from the environment and other objects through sensors and then creatively applies them in the process of self-calculation. Cognitive ability is also the building block of rationality.

Of course, only having cognitive ability is far from sufficient. Passively obeying the goals set by the programmer makes AI one of the ordinary tools without moral subject status. AI needs to set action goals for itself; that is, it should have "intent capacity". The so-called intent capacity refers to behavioral competence or capacity for responsibility, which means "man has the ability to independently and responsibly determine his existence and relationships within a given range of possibilities, set goals for himself and limit his behaviors according to his essential attributes". This ability includes communication, prudence, choice, decision and free will.

Although the levels of intelligence and cognition of AI on this occasion are no different from those of human beings, due to the lack of ethical regulations, the behaviors of AI can hardly center around the rights and interests of human beings, which may cause serious harm to human social order, core interests and even survival. On this proposition, we can cite a principle put forward by Finnis: "if the potential risks of damaging public good arise from the behaviors of our compatriots, we must accept them; but if they are caused by life beings other than our compatriots, then we don't need to accept them".

This requires AI to have a human-like moral ability and to try to reduce or eliminate the risk of harm to human racial interests as much as possible. Moral ability refers to the ability to judge and control one's behavior in strict accordance with moral practice under the framework of moral norms and ethical regulations. The premise for someone to have rights is that he is essentially a person with ethical significance. If anyone wants to obtain moral subject status relative to others, that is to say, he has the right to ask others to respect his own personality and has the obligation to respect others' personalities. Subjects with moral ability have sufficient moral reasons to ask others to respect their legal personality, while those without moral ability have no capacity to respect others' personality or the need to be respected by others.

Since the proposal of AI in the 1950s, AI still remains in a weak stage after half a century of development. The technology of AI is mainly based on the promotion of deep learning, big data applied in specific application scenarios and computing power is in the primary stage. It takes a large number of data elements as the basis for thinking, decision-making and action. Static AI based on big data belongs to weak AI. Its algorithms may achieve exponential performance improvement in virtual environments, but it can only realize linear performance improvement in real data environments, which has certain limitations.

From the technical level, the next generation of AI may move forward in the following three directions:

First, from the perspective of the underlying technology paradigm, AI technology with complex computing ability based on dynamic and real-time environmental changes could be a development direction, including big data intelligence, swarm intelligence, cross media intelligence, hybrid augmented intelligence and intelligent unmanned systems. These are all directions that are being promoted in China as the new generation of AI technology. In particular, the last two, which are implemented through a new bottom intelligent system, are particularly noteworthy.

The second direction is AI technology based on a new application paradigm formed under complex goals from the perspective of application target scenarios. The complex goal here is to build an artificial intelligence technology for an intelligent society with urban groups as the goal, which involves new technical directions such as the landing of urban full-dimensional intelligent perceptual reasoning engines and autonomous learning oriented to media perception. The typical examples are the "Woven City" initiated by Toyota in Japan and the future city "Neom" being promoted by Saudi Arabia.

The third direction is a revolutionary AI technology based on the new chip architecture. In the Von Neumann architecture, the mainstream architecture of AI chips, there is a ceiling of memory performance due to separated units of computing and memory. The result also leads to limited growth space for processor performance and affects the effectiveness of the computing power of the current neural network model for complex scenario processing. However, the adoption of in-memory computing in the new chip architecture is of great value. Therefore, international academic conferences on semiconductors such as ISSCC or semiconductor manufacturers such as IBM and TSMC are committed to promoting and implementing the innovation of in-memory computing architecture.

We can see that although the so-called "strong AI" with "free will" has become a hot topic among the supporters of "subjectivity", the following three characteristics of the current development of AI have forced us to doubt that scholars holding this view have fallen into a state of "utopianism".

In terms of research direction, the research on AI has not explicitly moved toward the field of strong AI, and strong AI is still an illusory direction at present and will not significantly improve the value and benefits of AI to mankind. In terms of technical architecture, neither the computing architecture based on semiconductor materials nor the algorithm development based on binary logic can realize the "independent evolution" of AI. To realize strong AI, human beings need to break through the shackles of existing logic circuits and data structures, which is almost impossible in the short term. The lack of a sufficient theoretical basis is another barrier. In terms of research principles, both scientific research and industrial development need to strictly abide by ethical and moral principles. Even if strong AI can be realized, its risks are far greater than the benefits. The Pandora box should not be opened.

Artificial intelligence is a factitious nonliving entity based on intelligent technology rather than life metabolism. First, AI lacks the physiological basis of the human "carbon-based" brain. The human brain is the core of human life. Brain death marks the loss of legal personality. The "silicon-based" AI program or robot does not meet the physiological requirement of subject personality based on the existing moral framework. Second, AI does not have rationality and will. Rationality covers people's perception of the characteristics and laws of the things around, the understanding of and compliance with ethics, and the empathy of emotion. Kant proposed that "human beings are rational and themselves are ends". Hegel agreed that "man has the purpose because of rationality", man has the ability of introspection and self-examination because of rationality. Different from human rationality, the "rationality" of AI is essentially a logical relationship based on an algorithm model, let alone human "natural" rationality. The value of its "rationality" is also to realize human will and purpose and much less introspection and self-examination, so it is impossible to break away from the restrictions of human regulation from the algorithm level for critical reflection. Although AI has deep learning ability, algorithm-based correction is essentially a kind of pattern recognition, rather than truly having natural rationality. The "intelligence" is a representation of algorithm simulation given by human beings.

Will is not simple logical calculus, and it contains two key factors: desire and action. Driven by desires, human beings make judgments on the ethical value of realistic propositions they face. Humans' desire and pursuit for survival, quality of life, identity and personality drive them to constantly transform themselves and nature to create a better living environment or make the most advantageous judgment. Neither AI has desire of its own and public nor emotional fetters. In the face of similar propositions such as the "trolley problem", it is difficult to evaluate the rationality of the decisions made by AI. In most cases, AI only performs probability calculus through human algorithms, let alone the possibility of establishing free will without the help of human beings.

Some critics believe that this idea of regarding AI as a person with nonfull capacity is full of discrimination against special populations who lack rationality and will as well as protection in human society, which is essentially an irrational embodiment. Moreover, children will also have rationality and will in adulthood. Vegetables with brain damage may also come to themselves and regain rationality and will, which obviously contradicts the characteristics of AI.

It should be emphasized that AI cannot bear responsibility under the existing moral framework, and giving it subject status is only a formal regulation. However, there are many gaps and blanks in propositions such as whether the control of AI on property and the ownership of property should go to AI itself or the human subject who makes it. Theories such as "finite personality" have essentially proved that it is unnecessary to endow AI with the qualification of moral subject, and the practice cannot pass the test of "Occam's razor principle".

The above is our preliminary thinking on the moral responsibility of AI. Later, we will discuss AI moral and philosophical propositions in various fields. The basic starting point for us to pay attention to these propositions in this book is that they are closely related to the fate of human civilization. As Will Durant and Ariel Durant defined in the last volume of their masterpiece, *The Story of Civilization*: civilization is social order promoting cultural creation. Four elements constitute it: "economic provision, political organization, moral traditions, and the pursuit of knowledge and the arts. It begins where chaos and insecurity end. When fear is overcome, curiosity and constructiveness are free, and man passes by natural impulse toward the understanding and embellishment of life".

In brief, only when we overcome the moral fear of AI in future human civilization can we create a brighter civilization. Human civilization pushes us to find solutions, and what we do is to find a way for humans and AI to achieve long-term coexistence while promoting civilization.

Chapter 3
Digintelligence Risk Society is Around the Corner

Mankind has accelerated into the "digintelligence society". Today, precision manufactured robots are busy working in Foxconn factories. Baidu Apollo driverless cars are carrying passengers on the streets of Beijing. The ubiquitous face recognition temperature measurement systems are working in office buildings. Some magazines like *Nature* are continuously reporting that AI outperforms doctors in the accuracy of disease diagnoses… These once incredible scenes are imperceptibly being implanted into all aspects of human life. According to the report of the Chinese Business News, Stanford University experts predict that 48% of American jobs may be replaced by AI in the future, while the percentage might be up to 70% in China.

Meanwhile, as big data science and technology are developing rapidly, data centers provide key infrastructure for all walks of life. The "big data development strategy" competitively developed by many countries also helps big data gain a critical position in philosophy. As a new philosophical theory, "dataism" may become one of the most important philosophical theories in the era of digintelligence.

The era of digintelligence, as the name suggests, refers to an era in which digitalization and intelligence coexist.

To make the concept clear, first, we need to look at the scientific and technological revolutions since modern times over a long period of thinking. In the dimension of time, the digintelligence era is staying in the transition period from the third to the fourth scientific revolution. At this stage, the leading technology of the third scientific revolution—Internet technology—has grown mature, and the market has become saturated. By virtue of transactions and interactions, a large number of Internet enterprises have generated and accumulated a large amount of data, powerful computing power and a large number of rigorous algorithms, which provides multiple possibilities for AI to predict the behaviors of consumers and producers. These Internet industrial revolution heritages have brewed and opened a new era of digitization and intelligence (digintelligence era for short). Moving forward from the traditional Internet era to the digital intelligence era is a process of creative destruction. The digintelligence era is not only a technical revolution but also an industrial revolution, life revolution and talent revolution. It is a profound and systematic wave of creative

© The Author(s), under exclusive license to Springer Nature Singapore Pte Ltd. 2022
Z. Liu and Y. Zheng, *AI Ethics and Governance*,
https://doi.org/10.1007/978-981-19-2531-3_3

destruction involving the economy, society and education. The fundamental force behind the digintelligence revolution in fields such as economy and society is the highly coupled result of digintelligence technology and entrepreneurial power.

In today's society, a force of globalization is developing rapidly and constantly shaping the world we live in. Globalization includes not only economic globalization, cultural globalization and technological globalization but also a kind of risk globalization. Under the background of globalization, human society is facing more risks than ever before, such as the risk of large-scale unemployment, the risk of widening the gap between the rich and the poor, ecological risks and so on. In 1986, Ulrich Beck, a famous German sociologist, first put forward the concept of "risk society" in his book *Risk Society*. He pointed out that the central topic revolved around the concept of industrial society from the sense of Marx and Weber is how socially produced wealth could be distributed in a socially unequal and "legitimate" way in a society of scarcity, to which the "risk society" is based on the solution. How can the risks and hazards systematically produced as parts of modernization be prevented, minimized, or channeled?

Based on the era of digintelligence, the "risk society" generates a deeper meaning. With the continuous growth of the digital economy, the degree to which humans transform social life and nature through intelligent technology continues to deepen. In a sense, with the help of digital technology and means, human beings have improved their abilities to deal with all kinds of risks, but they are still facing new types of risks brought by technologies, namely, technical and institutionalized risks.

Technical risk refers to the negative impacts of new technologies, including challenges to human rights, threats to system security, disclosure of privacy, etc. Institutionalized risk refers to the cumulative risks of failure of the system and normative frameworks that balance the contradictions of all parties in society under the operative rules in the new digintelligence era. In addition, many rules are waiting for us to recustomize in this transition period when new systems and old systems overlap.

Technical risk and institutionalized risk are the main types of risk structures in the digintelligence age and have global impacts. These risks have raised global risk awareness. Human beings need to have an overall identity in recognizing and handling risks, which has also given birth to a series of era concepts closely linking human destiny such as "human community with a shared future". Before clarifying our action plans against risks, we might also consider the following issues together:

First, is an algorithm the only way for organisms to develop? Is life truly just a process of data processing?

Second, what is more valuable, intelligence or consciousness?

Third, if unconscious algorithms with high intelligence know more about us than ourselves, what would this mean for society, politics and daily life?

To explore these issues, we have to essentially return to thinking about how to consider the value and risk boundary of technological development and determine the logic behind the risks. Although there are still many uncertain factors in a series of risks brought by AI and other technologies, what we are certain is that AI must reflect human needs and interests, including both the necessity of survival and the needs of development, spirit and society. All these needs relating to human measures of values and purposes continue to drive us to develop AI.

3.1 AI Ethics in Digintelligence Risk Society

The development of AI technology mainly takes computers as the carrier to promote the development of automation technology. AI technology based on data can help us with different sorts of production activities through various automatic devices, improving the overall development efficiency of society. However, great efficiency improvement and enormous risks accompany AI development. We need to think carefully. What kind of future will AI bring us? How do the risks of AI arise? How can relevant risks be prevented?

Let us first review the following events: in May 1997, the computer "Deep Blue" developed by IBM beat Garry Kasparov, an international chess master, for the first time. AI Alphago defeated Lee Se-dol, a top go master in South Korea in March 2016.

On October 25, 2017, Sophia, the world's first robot citizen, became the first female robot to be granted the legal citizenship and nationality of Saudi Arabia.

On November 7, 2018, the world's first complete model simulation intelligent synthesis anchor jointly developed by the Sogou and Xinhua News Agency officially appeared at the fifth World Internet Conference (WIC).

The chain of events has deepened people's concern and thinking about AI. As general AI technology continues to develop, people gradually and consciously accept robots into life and even regard AI products as their equals. Since the AI partner depicted in the film *Her* is gradually becoming a reality, we have to attach importance to the risks of AI development. We can discuss it from the following three aspects: **the risk category of AI, the formation mechanism of AI risks, and the boundary of AI development.**

For the risk category of AI, we still need to start with Ulrich Beck's concept of the "risk society" mentioned above. Beck believes that in a risk society, material scarcity has been replaced by risks, and they become the focus of social and political issues. Different from traditional society, the risks of modern society originate from the aggressive promotion of science and technology without limits, resulting in uncertainties in purposes and results. In other words, technology risk is not only the internal attribute of technology but also the result of human selection, which is fundamental to our understanding of technical risks.

Significantly, although the risks brought by AI belong to technical risks, they are quite different from traditional ones. Generally, technical risks come from external factors, including environmental risk, ecological risk, economic risk, and social risk. That is, risks are brought by the interaction between technologies and social factors. However, AI technology will bring great internal risks, namely, challenges to the status of human existence and the complexity of human boundaries and scales.

The risks of AI mainly include two parts: one is objective reality at the physical level, that is, gradual replacement of human abilities increases external technical risks, such as the sharp rise in the unemployment rate; the other is the risks at the psychological level of human beings. For instance, as robots gradually obtain human figure and cognition, humans begin to acknowledge that they are of the same homogeneous group. This inevitably raises ethical problems. From the perspective of the risk society, we can further understand that the risk society is the consequence of modernity, while AI ethics is the technical embodiment of the risk society. From another point of view, the social presentation of AI ethics is one of the signs of the arrival of the "digintelligence risk society".

If the risk society is the consequence of one-sided pursuit of maximum economic interests in the industrial society and the immoderate exploitation of natural resources, the digintelligence risk society is promoted by the excessive pursuit of the combination of technology and capital. The digintelligence risk society is an aggregation of human endogenous risks. It is precisely because human beings re-examine the relationship between themselves and machines that digintelligence social risks arise.

We will mainly look at the basic logic of AI risk formation in the following section. The biggest difference between AI and other early human technologies is the automation of behavior. AI systems can run without human control or supervision. Thus, it can be seen that the benefits and risks of AI technology to human society exist side by side.

On the one hand, AI liberates productivity by replacing human labor and gives human activities more time and space. On the other hand, AI puts human beings at the risk of losing control, and human subject status is allocated to machines to some extent. More profoundly, the potential driving force of modern technology development is humans' immoderate pursuit of technology. Humans have improved their capacities through AI technology. AI technology is implementing humans' willingness to expand technology by integrating a variety of technologies. At the same time, humans may gradually lose themselves in technology expansion, and technology risks are coming true as the development of technology departs from humans' will.

Apart from the characteristics of technology itself, it is easy to ignore that with another element of the technical revolution—capital—the development of modern technology since the industrial revolution can never eliminate the power of capital. Capital is profit-seeking, and the ultimate goal of technology is to realize people's pursuit of material interests. Therefore, technology can maximize the goal of capital. In other words, technology and capital reflect a certain degree of isomorphism. Capital can promote the advantages of technology in the process of value growth.

In short, technology can produce a force beyond human control. Like riding a certain vehicle, driven by technology, people exploit nature and plunder resources intemperately. In the meantime, human beings are firmly controlled by technology since they invest themselves in technology as a resource, which is not only the essence of modern technology discussed by Heidegger but also an important viewpoint to understand AI risks. Meanwhile, as capital has a kind of surreality that is not limited to any affair, it can change things according to people's needs. More importantly, since the development of AI technology will widen the digital divide, it is extremely important to pay attention to the social equity brought by digitization in this process.

Upon understanding the risks brought by AI, we will discuss the boundaries of the development of AI technology. Only by clarifying the boundaries can we understand its risks. Compared with the usual technical risks, the risks of AI mainly lie in the divergences of the understandings of the intelligence boundary. As humans are the only entities with intelligence, so-called intelligent boundary problems arise after AI emerges.

We can understand the concept of an "intelligence boundary" from the following three aspects.

First, the "intelligence" of AI is developed by converting information into knowledge through computer technology, so there is a computable boundary. Currently, computers focus more on logic-related operations, while emotion cannot be calculated. Of course, we also see that entrepreneurs such as Elon Musk interpret human thinking through brain-computer interfaces and other technologies and give AI a new intelligent mode. However, given that our cognition of the human brain (especially the part related to consciousness and emotion) is very superficial, we are not able to see the prototype of AI producing self-consciousness and concepts for the time being. If the advancement of technology can promote AI to generate consciousness, systematic risks will undoubtedly arise. Human beings have to coexist with them, and living space conflicts are inevitable.

Second, the boundary of AI technology challenges the boundary of the social attributes of technology. Generally, technology has natural and social attributes. The so-called natural attribute is the internal reason for technology to emerge and exist. In other words, technology conforms to certain physical laws. The social attributes of technology mean that technology should conform to the laws of society. We can see that AI still exists as a functional individual or group; since it does not possess the abovementioned social nature, it cannot be regarded as a species or a group.

Third, it can be seen that AI technology may promote the emergence of the "post-human" era; that is, it can exist through species different from humans, such as "cyborg". Cyborg has broken the boundary between the subject and the external environment, as well as the boundary between nature and society, and represented a concept of the "post-human era". In this concept, humans are constantly objectified, while objects are constantly humanized, and the boundary between human and thing is continuously blurring, thus forming a deep union between organisms and objects and giving the natural body the attributes of a machine. We can see that most of the "electronic men" or "artificial men" in the science fiction films are created based on the concept of cyborg, which allows human beings to surpass their limits endowed by nature.

In the movie "*Alita: Battle Angel*", the heroine only has the human brain, and the rest of her body is composed of machines with more powerful performance and skills. If this scenario happens in reality, it will cause two results: on the one hand, the vulnerability of humans as carbon-based creatures will be reversed, and we humans will have stronger survivability and gain new advantages in competition with machines; on the other hand, the human body has lost its natural uniqueness and become a replaceable and inorganic part.

Furthermore, the digintelligence risk society is a "real virtual" scene; that is, the risks come from uncertainties and bring a threatening future. After recognizing this threat, people began to have conflicting ideas. Such risks are also man-made results of the loss of the dualistic property of nature and culture. The consequence requires us not only to fully regard the problem of AI as a fact statement but also a value statement. We should not only use instrumental reason to change rules but also value judgment to constrain ethical bottom lines and regularize responsibilities.

We provide a solution to AI risks here. **AI must reflect human needs and interests**, which is also the basic starting point of AI research. This includes a wide range of needs, including survival, development, spirit and society. Meeting the core needs of mankind is the driving force for us to develop AI, that is, people's value scale and purpose. Human beings need to bear the responsibilities of science and technology in this process.

Master Qian Mu once said life evolves into human beings, and the biggest difference between human life and the lives of other creatures is that we pursue purposes other than survival. These purposes go beyond the purpose of survival, that is, the value of life and death as well as the brilliance of rationality. What we see in the development of AI is the maximization of human utilitarian purposes, which will consolidate the survival foundation of the whole human society. However, it lacks the practice of super utilitarian goals such as rationality and aesthetics and involves little content of human freedom and values. If we simply consider utilitarianism, undoubtedly, individuals and group organizations will become alienated. Human beings may develop self-denial, and their dignity and existence will also be challenged.

We need to realize that **the boundary of technology is an ethical problem**. There are internal contradictions between the possibilities of technology and the constraints of ethics. It is easy for us to only notice the power and role of machines and ignore the value of human beings in the process of technological development. Humans are in a passive position in the automatic machine system. Technologies not only represent the extension of human essence but also suppress human essence. If human life becomes the object of technological transformation, the technicalization of humans is inevitable. In other words, human beings are supposed to be manufactured. Technology will not only change "impossible" into "possible" but also "can" into "should".

We should see the internal contradictions between human life value and technology. Namely, human life is natural, while technology is man-made. Life is not recovered, but technology can be replicated. Human value is the criterion of human existence, and human beings need to realize the unity of emotion and rationality based on their own moral nature, and unity brings the highest orientation of human value. The development of AI technology must take humans and human social groups as the value scale from the internal and external perspectives, respectively. Taking humans as the value yardstick guarantees human subject status and independent consciousness, and taking human social groups as the value scale ensures the sustainable development of the whole society. AI will not easily overstep the ethical threshold and cause irreparable risks and losses only if it develops within the humanistic framework. This is not only the basic principle for us to understand the future development of AI but also the essence of our understanding of AI risks.

In conclusion, the AI ethical risks we have proposed are actually the core issues in the digintelligence risk society. When rationality transforms into instrumental rationality, human society becomes risky. The process of digitization increases deeper risks, that is, the arrival of "human autognosis risk". When the science and technology extended by knowledge can help human beings solve all problems, they only pay attention to power rather than human freedom and natural life with this technical knowledge, thus losing their sense of meaning and responsibility, which is the biggest crisis in the digintelligence risk society. We need to change people's wrong ideas and establish the correct concept of "tech for social good". The key is the establishment of ethical responsibility and understanding the essence of the social risks of AI.

3.2 Will Winter Come?

The AAAI Conference on Artificial Intelligence is one of the three comprehensive top conferences of international artificial intelligence. In 2019, the number of registered papers exceeded 10,000, and the number of valid contributions was nearly 9,000, compared with approximately 2,000–3,000 in normal years. Similar situations also occurred in the contributions of several other important academic conferences (such as CVPR, ICCV and ECCV). Among them, China contributed the largest increment. This upsurge in international AI academic circles reminds us of the condition in the second wave of AI in the 1980s. Unfortunately, the wave was soon followed by a cold winter. Thus, worries about the cold winter of the third round of AI have never stopped. To judge this problem, we must determine where the risks of the AI industry come from.

As for AI risks, the most sensational one is about the social division of labor and the change of social stratum. Some researchers believe that AI will replace some middle-class jobs in the future, which will affect the mobility of wealth and the distribution of social resources, making social wealth too concentrated in the elite class or hegemonic countries at the same time. With the help of automatic weapons and military systems, competitive barriers are infinitely raised. Others worry that AI, especially strong AI with self-awareness, will threaten human survival, go against the ultimate goals of humans, and ignore human instructions. Physicists such as Stephen Hawking have issued similar warnings, even if Elon Musk, the founder of the rocket company Space X and the billionaire entrepreneur of electric car manufacturer Tesla, holds a similar view.

Musk warned: "With artificial intelligence we are summoning the demon". Although Elon Musk's Tesla automobiles are considered to be the world's most successful autonomous vehicle products, the risk that overuse of AI may make it too powerful to control in the future is still a headache for him.

In fact, the problems brought by AI are not new. The birth and development of AI are accompanied by fears of AI replacing human beings in some work, benefiting only a tiny minority, or even subverting the whole social order. Even in Britain, where industrialization grew vigorously two hundred years ago, these problems were already vexed. However, at that time, they were classified as "machine problems" rather than "industrial revolutions". As early as 1821, economist David Ricardo worried that machines might have a far-reaching impact on all sectors of society, especially the working class, causing damage to their interests. Thomas Carlyle, a historian, criticized machine as a "mechanical devil" in 1838. He held that machine would become a destructive product that would "disturb the normal life of all workers".

The Bank of America Merrill Lynch predicts that by 2025, the annual cost of innovative destruction caused by AI will reach 14 to 33 trillion US dollars, including 9 trillion labor costs for its ability to make knowledge work automatic, 8 trillion production and medical costs, and 2 trillion efficiency gain brought by the use of driverless cars and UAVs. McKinsey Global Institute, a think tank, claimed that AI is promoting social transformation, and the transformation is "10 times faster, 300 times larger in scale and almost 3,000 times greater in influence" compared with the industrial revolution.

As the story of AI is being hyped so mysterious at the moment, we need to return to the origin of deep learning technology to see what has happened.

The concept of AI was officially established at an academic conference of Dartmouth College in 1956. Scholars hope to define its tasks and development path by identifying AI as an independent science. They declared that many features of AI can be described by accurate logic operations and can be simulated and realized through machines. At this conference, the term "artificial intelligence" was mentioned and used for the first time, and it then kept developing spirally along the two paths, symbolic computing and neural network, in the following decades. However, at that time, people were often unduly optimistic about AI. Although the so-called "expert system" or "neural networks" have achieved some phased results, many descriptions of AI still cannot be fulfilled. In 2012, an online competition called the "ImageNet challenge" ignited public passion for AI.

Deep learning technology can be understood as the "integration of the past fragments". Yoshua Bengio, a computer scientist at the University of Montreal, one of the pioneers of deep learning, believes that, in essence, the use of deep learning based on powerful computing power and massive training data to strengthen the development of AI stems from an old idea—the so-called artificial neural network (ANN).

Deep learning can be divided into different categories. One of the most widely used is "supervised learning", which uses labeled instances for system training. For example, in the case of spam filtering, a large database can be formed by a large number of sample emails that are collected and then marked as "spam" or "nonspam". A deep learning system can use the database for training and improve the accuracy of the spam check by continuously inputting these examples and adjusting the weight in the neural network. The greatest advantage of this method is that it does not require human experts to establish a set of rules, nor does it require programmers to type the code to execute the programs. Systems themselves can learn from labeled data.

3.2 Will Winter Come?

A large amount of data can support supervised learning. The frequent use of this technology has prompted some companies previously engaged in financial services, computer security or marketing to reposition themselves as artificial intelligence companies.

Another technique, unsupervised learning, also exposes neural networks to a large number of examples for training without telling the learning mode for which to look. The neural network will automatically identify features and cluster similar instances to reveal hidden clusters, associations or patterns in the data. Unsupervised learning can be used to search for things you do not know exactly what they look like, for instance, monitoring network traffic patterns to prevent cyber attacks or checking large insurance claims to identify new types of fraud. Here is a best famous case. Andrew Ng led a project called Google Brain, a huge unsupervised learning system for identifying common patterns in thousands of unmarked YouTube videos, when he worked at Google in 2011.

In addition to the above two kinds of deep learning, there is another technology named reinforcement learning, which lies in between. Guided by the concept of reinforcement learning, the neural network trains the model through continuous interaction with the environment, and the environment will give feedback in the form of rewards. The center of reinforcement learning is to continuously adjust the weight in the neural network through training to find a sustainable strategy that can produce higher rewards.

From the current situation, Google, Facebook, Microsoft, IBM, Amazon, Baidu and other companies have opened the source code of their DL software for free download. The reason they do this is that the researchers of these companies want to make their work public to search for more potential talent. Meanwhile, by opening AI software to developers free of charge, these manufacturers can obtain the basic element—a large amount of data—for large-scale training.

After understanding the origin of this round of AI technology, we can obtain three conclusions. Currently, the development of AI technology is based on three foundations.

(1) Effective deep models. Basically, deep neural networks derived from the logic of mathematical statistics are widely used at the stage, so there are problems of boundary and interpretability of technology application scenarios;
(2) Strong supervision information. All data should be labeled, and the more accurate the better. Thus, we have problems with efficiency and costs.
(3) Relatively stable learning environment. Robustness is another problem. Once the learning environment becomes unstable, there will be considerable deviation in the output results.

It can be seen that the reality of this round of AI technology development is not as "sexy" as we think. This round of development mainly comes from the progress of hardware (bringing more memory and faster computing speed) and the growth of data (bringing the improvement of training efficiency). Today's AI can be understood as "narrow AI" that can only be applied to the specific contexts for which they are designed. Such a situation makes our imagination of AI much empty and helps us orient our efforts in the future.

Academic discussions on the above risks mainly focus on the fields of robustness, efficiency and interpretability. Actually, in the last few years of research, academia in China has begun to pay attention to the study of "closure". Its basic logic is that the current deep learning is mainly created by adopting the thinking models of "brute force" and "training algorithm". The basic idea of the former is to establish knowledge representation according to the accurate model of the problem and enumerate the possible solutions to the problem by reasoning or searching in compressed space to obtain the optimal solution. The basic idea of the latter is to establish the meta-model of the problem and then label and train the data according to the meta-model, select the appropriate neural network structure and learning algorithm for deep learning, and finally obtain the corresponding learning model by data fitting. On the basis of the understanding of the idea, domestic scholars represented by Chen Xiaoping have begun the study of closure criteria. To put it another way, whether a practical problem has the property of closure or can be closed can determine whether AI technology can be applied to a specific field in principle.

In summary, this section discusses the development history of deep learning and the risks concerned by the AI community, and it also makes us understand that AI technology represented by deep learning has boundaries. We recognize a basic fact through thinking. The existing AI technology can be applied on a large scale, and what we need to do is to solve the corresponding scalable problems in different industrial practices and avoid the relevant industrial, technical and ethical risks.

3.3 Data Privacy and Social Responsibility

Data privacy protection is a typical case of undertaking social responsibility. In fact, "responsibility" is not only a legal concept but also constitutes the basis for people to understand moral obligations. Responsibility actually goes beyond what we usually call morality. We should guide people through the AI ethical challenges of the risk society and point out a clear direction of ethical responsibility through the concept of responsibility.

Let us look at specific data ethics issues. The existence of data elements is the foundation of the existence of the digintelligence society, and it is also one of the most vital propositions in AI ethics. In other words, we are in superposition of the real world and the data world. Fundamentally, the problem of data ethics is not just the protection of a single right but the cognitive reconstruction of the operation logic of the digital world. Among the many recent discussions on data privacy protection, there is little bottom-level discussion related to the basic logic of data privacy protection. Many discussions focus on the importance of "data privacy" and the concern over the risks of data abuse. Therefore, most discussions remain in the controversies on public issues and the superficial discussion of scientific and technological ethics.

3.3 Data Privacy and Social Responsibility

Two fundamental issues are involved in this topic: first, the essence of data value as a factor of production in the whole intelligent era, including how data will be used after data mining and how the paradigms generated from new knowledge affect the real world. Second, what is the core purpose of the existing concept of data privacy protection, and how can we understand the relationship between technology and humans behind data?

The core of the first question is "value", including the meaning and measurement of value. In the discipline of digital economics developed by the author, "value theory" is placed at the center of the whole discipline. The author believes that the center of data value theory is to put the value proposition of "innovation-driven development" at the core to realize the internalization of information productivity and the production mode. Since Adam Smith opened the history of modern economic science, the two factions, labor axiology and exchange axiology, have been debated for more than 200 years. From the perspective of digital economics, labor is indeed the basis of value creation, but without a socialized division of labor and cooperation, value creation is impossible. The same applies to the data value. It is related not only to the behavioral value of individuals in the digital space but also to the value created by data in collaboration with others in the community. Therefore, in the study of data value, we introduce the concept of "hierarchy", which is also an important technical concept in complexity research. We assume that the complexity in the digital economy system comes from the emergence of new subjects, phenomena and laws after individuals make up the whole. When individuals form new interest subjects, new markets and new networks in the digital space, new values are created.

The core of the second question is "responsibility", which includes the embodiment of individual responsibility, corporate responsibility and social responsibility in data privacy and ethical issues. The biggest problem faced by individuals serving as the subject of data rights is that enterprises and social organizations excessively use personal data, resulting in the loss or even deprivation of a series of rights. The rights to data privacy and data fairness represented by "algorithm discrimination" are the most central rights among them. The typical phenomenon of the former is that tech enterprises are realizing their interests by virtue of user data abuse, while the representative phenomenon of the latter is the unequal opportunity brought by the improper use of algorithms in various social organizations. Therefore, how to make individuals, enterprises and society realize their responsibilities in the process of data use and the equivalence of their rights and obligations in digital space without losing their commitments to data responsibilities due to the existence of technology is the key to all data privacy protection.

Historically, the emergence of privacy has changed the relationship between man and technology and brought a new perspective. As early as 1890, American Warren and Brandeis pointed out the problem of privacy and held that new inventions and commercial methods trigger risks. In 2006, Irving shermilinsky definitely pointed out that we should re-examine privacy and believed that we should further establish the legal responsibility for digging deep into personal information. To discuss issues related to data privacy protection, the *GDPR* approved by the EU in 2016 is a document that must be mentioned. In reality, the document is generated on the basis of the *Data Protection Directive* adopted in 1995 by the EU to regulate the acts of processing personal data within the EU and an important part of the privacy law and law of human rights of the EU.

We can see the changes in the EU's data governance thinking compared with 20 years ago. If the *Data Protection Directive* is a normative document that governs the automatic decision-making of machines based on data privacy protection, the legislative purpose of the *GDPR* is not to concentrate on the use of machines but to completely reshape the relationship between humans and machines through the ways of conferring rights. In other words, the problem is how to establish a human-computer interaction relationship that meets the needs of the development of human society.

The data minimization principle and appropriateness principle proposed in the *GDPR* stipulate that the controller has the responsibility to take appropriate technical and organizational measures to ensure that only the personal data necessary for a specific processing purpose can be processed by default; otherwise, the personal data should be kept in a state of "simplification and locking". Different from the *Data Protection Directive,* which only emphasizes data and information security, the *GDPR* stresses the principle applicability of the whole chain from software or hardware design to business strategy and practice.

In regard to data privacy, the core idea of data ethics consideration presented in Article 25 of the *GDPR* is to design the concept of privacy protection into the corresponding software system and embed human ethical values into the process of technical design on the grounds of enhancing privacy data protection technology, which is also the core of solving future data privacy and ethical problems in the eyes of the author. That is, in addition to conservative methods of data privacy protection, the justice of behavior and the rationality of ethics can be embedded into AI ethics through technologies.

After understanding this point, we believe that the core of data privacy protection and algorithm decision-making lies in whether we can protect human rights through the paradigm designed by the program and implement the corresponding ethical values at the technical level. Of course, we can predict how many obstacles lie ahead, including the difficulty in converting the wording and grammar of legal language and clarifying the scope of complex and obscure legal rights and obligations, cross data privacy protection, algorithm transparency and so on. However, the path is almost the only way to achieve the balance between rights protection and social development in the context of technological development. Instead of falling into endless ethical disputes, it is better to consider how the technical path can realize such a complex system.

We are in a superimposed state of the real world and the data world. Fundamentally, the problem of data privacy is not just the protection of a single right but the cognitive reconstruction of the operation logic of the digital world. This brings about the particularity of data as a factor of production, which has changed the ontology, epistemology and ethical values of the existing world. From the perspective of ontology, the world in which human beings live has generated a new element—data, and unnatural products such as intelligent machines, thus forming a new living environment.

3.3 Data Privacy and Social Responsibility

From the viewpoint of technological philosophy, the trajectories of existence are presented in the form of data that bring about the change of "presence". In other words, human presence can realize the feature of expansion through digital technology and abandon its transparency. For example, human motion trajectories and time series can be reproduced through image recognition technology, and human behaviors can also be guided through data mining of shopping malls. This progress is changing the logic of human behavior as the most important value of data elements. From the angle of cognitive theory, the world faced by human beings is no longer a simple transformation relationship between human beings and nature but has generated new species and elements. Thus, the ways of perceiving the world and self-identity are also changing.

A series of recent news events on data privacy have deepened individuals' concerns about privacy. In the meantime, the complexity and opacity of the Internet economy relying on demographic dividends and the decision-making system based on machine science have amplified such concerns. The preconditions of how to give full play to the corresponding regulatory regulations are becoming harsh. Here, we must emphasize the fact that both algorithms and regulatory measures point to the "transparency" of algorithms. In fact, it is almost infeasible for the existing artificial intelligence algorithms based on deep learning. Moreover, the right to transparency does not necessarily ensure substantive competition or effective subsidy measures. It is more about giving the public a sense that regulators are doing something. In addition, actual technology and application scenarios have not been taken into consideration. Correspondingly, the *GDPR* only provides the stipulation of "processing personal data fairly and transparently" in principle, which provides a relatively open and flexible scale for the laws of data privacy protection to measure the impact of data privacy issues and make the algorithm decision-making system take responsibility.

To interpret the ethics of data privacy and algorithm decision-making, on the one hand, we recognize the importance of personal data rights and the necessity of establishing corresponding regulatory mechanisms; on the other hand, we should also reduce the occurrence of "normative disconnection" and "excessive extension of supervision". Privacy protection is not an excuse for inaction but a starting point for algorithm regulation. What we want to protect is the fairness and justice of data after it becomes the basis for algorithm decision-making.

In other words, we acknowledge the two-way regulatory effect of technology and ethics. On the one hand, we believe that technology constitutes freedom through the transformation of the material environment, in which there is a special element of data. Human beings are understanding such a data world and forming new ethical paradigms. On the other hand, technology and human beings form special connections that let us take positive and vigilant attitudes toward technological development. Meanwhile, they are also adopting technical adjustments to formulate new rules and open up new interpretation space for the moral development of human society. Namely, nonhuman elements (such as data) should be incorporated into the moral system.

Therefore, data privacy is not so much an issue of letting commercial rights cross the moral threshold. A deeper understanding is that human beings have to improve their technical abilities and revise the ethical paradigm to achieve a wider range of justice (including both the traditional world and the factitious world of data). We should pay attention not only to the ethical problems caused by technologies but also to the accelerating effect of technology on ethics.

The above is our practice and thinking about how to bear the ethical responsibility of the risk society in terms of data ethics. We can see that the three parts, individuals, others and society, are the responsibility subjects and objects of one another. Ethical responsibility and the need to shoulder them all come from the relationships among the existence of human beings. Responsibility exists among people, between people and society and between people and nature. It is an abstract existence that everyone and the organization can get through experience. It constitutes the moral foundation of individuals and organizations in the face of the risk society. In reality, the discussions on responsibility provide the meaning of understanding human life and the belief in world values, which has made it the center of all current moral arguments.

As Marx pointed out, "as a definite person, a real person, you have rules, missions, tasks and whether you realize this, it doesn't matter". Modernity has wrongly changed the evolution direction of human civilization and generated risks. The most important thing that human beings can do is to break through the shackles of modernity and completely revolutionize their own morality to get rid of the crises. The core of the goal is "responsibility". In regard to the specific field of AI ethics, the aims are to undertake the technical responsibilities brought by the information technology revolution, find the moral paradigms for human beings to get along with machines, and more essentially, shape a harmonious order to ensure that the risks of the digintelligence society are controllable, create a harmonious, safe, happy and healthy future of human society, and avoid human society moving toward the risks, conflicts and crises at realistic and moral levels. Everyone should have such a sense of responsibility and be responsible for others, nature and society.

Chapter 4
AI Medical Treatment: Epidemic, Death and Love

In 2020, COVID-19, a public health event that almost overturned the global landscape, threw human beings trapped by weak technical progress into unprecedented global crises at the start of the third ten years of the twenty-first century. Companies and factories were shut down, schools and airports were closed, global trade activities almost came to a halt, and various macroeconomic indicators repeatedly set new all-time lows. The epidemic has particularly impacted the basic epidemic prevention and emergency systems. A number of intractable and urgent challenges emerged, including a huge supply gap for protective materials in the early stage of the fight against the epidemic, the production capacity fluctuation of pharmaceutical and medical device enterprises, the blocked import and export of raw materials, etc.

Under the magnifying glass of the epidemic, the medical industry is facing an extremely severe test, and its recovery process is bound to be full of hardships. However, the reason mankind is great is that we always thrive in crises. Thousands of medical workers rushed to the front line for support. The construction of isolation hospitals and the development efficiency of therapeutic drugs and vaccines continued to show "human speed". The epidemic has promoted the emergence of many new medical needs that have driven the penetration and empowerment of intelligent technology into the medical industry in the digital era as catalysts and accelerators for the prosperity of the medical industry.

As early as the 1950s, John von Neumann, the "father of computer", pointed out that "technology is growing at an unprecedented rate… We will develop in a direction similar to the singularity. Once we surpass this singularity, the human society we know will become quite different". From then on, "singularity" was used by the academic community to refer to the moments when revolutionary changes triggered by technical changes occur in social life. Actually, the development of AI is on the verge of singularity, and it begins to play an increasingly important part in the medical industry.

From a positive point of view, AI can greatly alleviate a series of long-standing problems faced by public health. Meanwhile, AI is also likely to change humans' understanding of their own health and improve their lifestyles to a large extent. Therefore, with the help of AI, human beings are expected to solve many technical and social problems faced by public health, opening up broad new prospects for human health and medical treatment. The medical industry, including clinical medicine, medical research, biomedical research and development, medical equipment manufacturing, and medical waste disposal, is the major livelihood plan that directly affects human rights and interests of life and health as well as public welfare and is about building a moderately prosperous society in all respects and realizing the great rejuvenation of the Chinese nation. The epidemic has exposed many knotty problems faced by China's medical industry. As AI technology continues to develop, the demand and trend of energizing the medical treatment industry with AI technology are becoming increasingly apparent.

Medical AI applications such as medical robots and intelligent interrogation emerge one after another. The ensuing medical, technical, legal and ethical problems deserve further analyses and discussions when people weigh the loss of life and warm love. Two thousand years after Hippocratic Oath laid the foundation of medical ethics, the invention of AI may bring the greatest challenge to medical ethics in history. What will happen if an AI system makes a wrong decision? Who will be responsible for the problem that may occur? How can clinicians verify and even understand the content of the AI "black box"? How can they protect patient privacy? How can the biases of AI systems be avoided?

The chapter first analyzes the specific impacts of AI medical treatment on human social life and its development trend by reviewing the "black swan incidents"—epidemic and influenza in human history, and then discusses the social ethics propositions of AI and the value of moral machine.

4.1 Black Swan Incidents: Epidemic and Influenza

The novel coronavirus broke out all over the world in 2020, exerting a huge impact on human life and the economy. It can be said that the outbreak of COVID-19 has accelerated the influence of the great change unseen in the world in a century. If the former can be taken as the "black swan incident", then the latter can be regarded as "the gray rhino". Affected by the epidemic, the centenary change has accelerated evolution and fission in many aspects and promoted changes in China's relations with the rest of the world. New technologies represented by AI technology are playing their unique role during this process.

Over the last few hundred years, every change in the international order was spawned by a world war, such as the Westphalian System after the 30-year war in Europe, the Versailles-Washington System after the First World War, and the Yalta System after the Second World War. The basic outline of the current international order is mainly shaped after the Second World War. However, after more than 70 years, the world has gone through a series of events from the end of the cold war in 1991 to the "September 11 attacks" in 2001, the financial crisis in 2008 and Trump's

4.1 Black Swan Incidents: Epidemic and Influenza

victory in 2016, and the existing order has been grievously shaken. Although the "basic frame" still exists, the roles played by international organizations such as the UN, WTO, IMF, IBRD and WHO and international rules, including international arms control systems and driving force mechanisms for cooperation between major countries and international standards, have gradually declined. Moreover, America's ability and willingness to lead the world has also diminished. The international order is on the verge of collapse.

The outbreak and large-scale spread of COVID-19 are contributing to global gloomy conditions, including lockdown at the country level, stagnant economy, stock market volatility, plunging oil prices, interrupted communication, harsh words and rumors. The impacts and influence of this disaster are nothing short of a world war. The existing fragile international order is taking another hit. **The old order is hard to sustain, while the new order has not yet been established, which is the essential feature of the global change unseen in a century and the root of the current unpredictable and perplexing international landscape.**

Looking back on the past, since the 19th century, the improvement of the public health system and the progress of modern microbiology and medical technology have kept most infectious diseases under control. Mankind has achieved phased victories over the epidemic by relying on civilization and science. However, looking ahead, the rapid evolution of viruses, contact between humans and animals, the increase in urban scale and population density and the acceleration of population mobility will continue to bring about new epidemic situations, challenging governments' social governance and emergency epidemic prevention capacity as well as international cooperation mechanisms. We still have a long way to go to completely overcome epidemics.

From the perspective of human history, epidemics are a major disaster that can be juxtaposed with famine and war. Epidemics such as black death and smallpox have killed hundreds of millions of people, and historical tragedy has repeated itself repeatedly. In the face of the threat of the epidemic, human society has not only learned many painful lessons but also accumulated valuable experience and measures during the long-term fight against it. Civilization continues to move forward in one crisis after another.

We all know the influence of the black death in the 14th century on the European social system, but this is not the first terrible plague in the old continent. As early as 800 years ago, before the plague, the same disease named the Plague of Justinian spread rapidly in Europe and the Middle East in a similar way. The terrible plague lasted from 541 A.D. to approximately 750 A.D. It first appeared in Pelusium on the coast between Egypt and Palestine in July 541 A. and then spread to the nearby Gaza Strip in August and Alexandria, the capital of Egypt, in September. On March 1 of the following year, Justinian, the emperor of Eastern Rome, claimed that "deaths have been everywhere". The resulting disaster proved to be devastating. The death toll reached 5,000 every day, sometimes even over 10,000 or more. Almost all Byzantines experienced the pain of burying their loved ones, and some even secretly or forcibly threw the dead family members into others' graves. As a consequence, order became worthless in human nature and panic.

After that, what impressed human history most was the black death. In fact, its periodic recurrence lasted until the 17th century in Europe and the 19th century in the Middle East. This may be the most notorious plague in history, but it is by no means the only one. When it began to recede in Europe, the Spaniards crossed the Atlantic to the New World, bringing a plague of similar scale and even more devastating to the latter.

If the Spanish flu in 1918 promoted countries to establish a global cooperation mechanism on epidemic warning and control, make the public health system an indispensable key element in national governance and gradually modernize. COVID-19 has pushed the rapid development of digital technology, including AI, to infiltrate into the medical industry to achieve digital empowerment and transformation in the vertical domains.

Taking computer participation in drug screening as an example, technology giants IBM and Oak Ridge National Laboratory of the U.S. Department of Energy once said that they have screened more than 8,000 compounds with IBM's supercomputer Summit and has identified 77 small molecule drugs that may become potential therapies for COVID-19. In addition, Cyclica, a global biotechnology company headquartered in Toronto, reshaped the discovery process of the drug and shortened the preclinical pharmaceutical cycle with the help of AI, computational biophysics and biomolecular data. The company also assisted pharmaceutical companies in evaluating drug safety and effectiveness through multiple pharmacology and multitarget drug properties and jointly developed new drugs with laboratories and pharmaceutical companies around the world. Moreover, Cyclica, by virtue of the introduction of Z-Park River Capital (ZRC), one of the early investment institutions, and CCAA, has been working with CAMS to evaluate multiple potential drugs for novel coronavirus pneumonia from FDA-approved clinically safe drugs based on Cyclica's self-developed proteome screening engine MatchMaker.

The achievements of AI in the prevention and control of COVID-19 are also worth mentioning. In the prevention and control of infectious diseases of the new coronavirus in early 2020 in China, CT images served as an important basis for the diagnosis and evaluation of the disease. However, due to the large number of patients, many pulmonary lesions, rapid changes, and the need for multiple re-examination in a short time, imaging doctors face great challenges in accurate diagnoses and quantitative analyses. Taking the quantitative evaluation of CT images as a reference, most medical staff sketches the ROI manually, which is similar to manual stroke and matting in PS. It usually takes imaging doctors five or six hours to complete three or four hundred CT images for each patient. Generally, a patient needs to take CT images approximately four times from admission to discharge, thus resulting in an extremely heavy workload for radiologists. Doctors in epidemic areas may be able to determine patients' conditions from CT images in 5–10 min through a large number of cases. However, doctors in nonaffected areas often show hesitations in the process of diagnoses because of little experience in receiving relevant cases. They dare not make a diagnosis until the nucleic acid test is positive. The time spent hesitating and waiting may cause further cross-infection or even cluster transmission. Thanks to AI technology and CT images, as many pulmonary nodules can be detected as possible, and the time for the whole inspection process from image acquisition to result presentation can be shortened to seconds.

In fact, the impact of AI on medical treatment is not limited to drug screening and intelligent diagnosis outlined above. Other fields, such as precision medicine, health management and biotechnology, are also areas where AI can play a great role. From the above analysis, we can see that AI will enhance the development of the medical field toward intelligence, routinization and humanization more efficiently, especially the development of precision medicine. At the same time, these changes will also have an important impact on medical employment and humans' understandings of themselves.

In summary, the three positive roles of AI in future medical development that deserve attention are as follows:

First, human health care staff should be supplemented with AI "doctors" to alleviate the shortage of medical workers in the future. The training process of human doctors is quite complex, relatively costly and time consuming. Even in developed countries, the lack of experienced medical staff is a norm. With AI technology, countless "AI doctors" with required skills can be trained in a relatively short time, which can effectively solve the problem of insufficient human doctor resources. Of course, this does not mean that all human doctors will disappear in the future. In extremely complicated and creative cases, the role of human doctors is still irreplaceable.

Second, AI is helping increase the efficiency of drug mining and speed up drug development. The level and scale of drug research and development determine the development form of the medical industry to some extent. Thus far, the R&D of new drugs still quite costly, and even so, no one can ensure that it will work finally. The application of AI can greatly settle the corresponding matters. For instance, when screening new drugs, we can use the strategy network and evaluation network of AI as well as a Monte Carlo tree search to select the safest compound from thousands of alternative compounds as the best option for new drugs.

Third, AI may improve the level of reasonable application of medicine and finally cure cancer through precision medicine. With the help of big data and AI, accurate detection of lesions from each patient, tailored use of medicine, dressing change and medicine dispensing can all be realized, which greatly cut drug prices. Big data analyses based on AI can work out customized use of medicine suitable for individual conditions. Furthermore, AI and big data can also help change drugs for cancer patients through the detection of new lesions.

4.2 Breakthrough and Ethical Principle of AI Medical Treatment

After a basic understanding of the application of AI in the medical field, let us look at the ethical proposition of AI medical treatment. Compared with the technical optimism of AI in the medical field, an increasing number of people have begun to worry about its significant impacts on society.

First, patient privacy needs to be properly addressed. Big data gathering will inevitably involve the privacy of patients, and how to coordinate and balance the relationship between privacy protection and data acquisition is a significant problem faced by intelligent healthcare.

Second, social concepts and supervision should also be taken seriously. To gain the trust of the public, we need to effectively supervise the use of the big data and algorithms of AI healthcare and formulate corresponding laws and regulations.

Third, human bioethics also faces challenges. As mentioned above, human beings are expected to use AI to defeat cancer, but another more important domain is gene editing. Some technology giants hope to slow down or prevent human aging through AI technology. The realization of this goal needs to encounter many technical problems and insurmountable human ethical problems. If human beings obtain "immortality" to some extent, the ethics and rules of human society will undergo subversive changes.

Currently, AI ethics for medical treatment involve personal data and other aspects, which are closely related to each of us. **Next, the author will comprehensively deconstruct and reconstruct the ethical issues of the AI medical industry, mainly from the five perspectives of the public and patients, medical staff, public health institutions, AI medical business organizations and social management, striving to guide the public to construct a systematic thinking framework of AI medical ethics from different angles.**

1. **For patients and the public, privacy and security are the indicators of most significance**

Data privacy is the top priority that needs to be addressed. The development of medical AI technology and industry is inseparable from the support of massive data information. Both the most common application of AI in medical images and the ongoing electronic medical record and digital hospital management system that have been implemented in major hospitals take vast amounts of personal privacy data of patients as the training basis of pattern recognition.

Since precision medicine needs to formulate different diagnosis and treatment schemes according to individual differences, patients have no control over their private information. The "convenience", "efficiency" and "precision" enjoyed by patients are realized at the cost of handing over a wide range of private data, such as personal information, disease status, past medical history, family medical history, living environment, lifestyle, diet and daily life, clinical data, medical image data, and even genome data.

4.2 Breakthrough and Ethical Principle of AI Medical Treatment

Once the data are illegally leaked or used by criminals, some medical institutions driven by profits may sell the personal information of patients as commodities to other individuals or organizations, which will undoubtedly cause serious harm to patients. Banks and insurance institutions may arbitrarily raise the loan threshold or insurance premium once they obtain personal medical history. The employer may discriminate against or even refuse to employ a job seeker if they gain his/her personal health information and learn that he/she suffers from a chronic disease, although the chronic disease has no adverse impact on his/her work or life. Once the personal diagnosis and treatment information of the patient is leaked and obtained by advertising agencies, they may precisely push drug advertisements or medical institutions to the patient according to his/her disease type or health status. This condition is likely to mislead the patient into choosing medicine privately and going to illegal medical institutions instead of listening to the therapeutic measures of formal medical institutions, leaving the patient irreversible severe physical and mental trauma due to delay in diagnosis.

Security vulnerability is the key to the survival of the AI medical industry. Since the AI industry chain is intricate and complex, it is conceivable that there is a higher possibility of security vulnerabilities. For example, imagine that if a surgical robot is performing a highly difficult and accurate surgical operation, even if under the control of the doctor, any error may directly lead to the failure of the operation. In this case, neither the patient nor the doctor can determine who should be to blame for the failure. The ultimate person liable may be the "direct human error" of the doctor who controls the robot on site, the "indirect human error" caused by the robot manufacturer or algorithm provider, the subjective and malicious physical attacks or remote interference launched by external individuals or organizations, or accidents resulting from power or data transmission interruption. Sometimes such failure even has nothing to do with doctors, suppliers or manufacturers, it is just the consequence of a bug in algorithm…

2. **For medical staff, AI empowerment in clinical medicine threatens the dominant position of doctors and materializes the doctor–patient relationship**

As AI is gradually going deep into clinical diagnosis and treatment, its advantages make us unable to surpass and challenge the subjective position of doctors in the process of diagnosis and treatment. Every correct decision made by a doctor is based on solid professional knowledge and long-term experience accumulation, while the powerful computing power and information storage ability of AI make it possible to complete the study of all medical knowledge and clinical cases that an attending doctor may need decades of study in just a few seconds. AI medical treatment has higher accuracy and work efficiency than human doctors and will make fewer errors for not being restricted to emotion and energy. Such an efficient, accurate and convenient model not only greatly improves doctors' work efficiency, saves labor costs and optimizes work results but also threatens doctors' decision-making and contribution weight to the medical project itself and makes doctors suffer the threat and pressure of being replaced.

On the one hand, AI medical treatment challenges the subjective position of the doctor in medical diagnosis and treatment; on the other hand, it may lead to doctors' excessive dependence on AI technology in the future and continuously falling requirements for medical professional knowledge learning. Meanwhile, as AI gradually penetrates into clinical diagnosis and treatment, the participation rate of doctors in the process of diagnosis and treatment is declining, making it increasingly harder for doctors to establish effective communication and stable relationships with patients. However, medical diagnosis and treatment include not only test data, the collision of science and experience but also direct communication between doctors and patients. Patients obtain psychological and social support to fight diseases and unknown anxiety and panic from doctors. Doctors will provide appropriate intervention according to individual psychological status to alleviate patients' negative emotions and improve the effect of diagnosis and treatment. This implicit spiritual needs and psychological support based on the effective communication between doctors and patients is sometimes more conducive to patients' recovery than drugs or surgeries, which also shows personality respect and humanistic care for both doctors and patients.

3. **For health institutions, research-based hospitals and scientific research institutions, data quality and platform stability are the basic indicators to ensure the high availability of AI medical treatment**

Let us start with data quality challenges. The large-scale and labeled data sets in the base layer are the foundation for the construction of the overall architecture of AI. The quality of the data directly determines the fitting degree of the AI model to the actual indicators. In AI medical treatment, machines can learn through an array of data sets involving medical textbooks, clinical cases and various physiological indexes of patients. How to obtain high-quality medical data sets is a knotty point faced by AI medical treatment.

Moreover, the stability of network attacks and AI platforms directly determines the success of medical and scientific research institutions and the stability of confidential data storage. Built on the huge computer and communication industry, AI medical treatment is bound to have a higher risk of being illegally stolen, attacked, modified and destroyed due to the complexity of its industry chain. Medical big data and AI platforms relying on cloud computing are examples. Although the local system integrates a reliable security firewall, they will become the targets of hacker attacks when the encryption measures of cloud storage or network transmission links are not up to standard. Moreover, AI also has many uncertainties. When AI medical technology enters a strong phase, that is, when it has human-like consciousness and perception ability, AI will fall into the dilemma between reason and perception in the case of complex and difficult cases. In moments like this, human beings often make final decisions based on moral standards or legal norms, but can moral cognition and the legal understanding of AI support them in making decisions that neither harm humankind nor fail?

Even weak AI limited by the human cognitive level has met a series of dilemmas of moral choice in the layout of AI moral algorithms, and the algorithm itself is likely to come across wrong instructions that have never encountered during the operation of the parameter adjustment test. Does AI have enough powerful and perfect error tolerance and correction mechanisms to ensure that the stability of the entire platform is not threatened in the event of incorrect instructions? With the continuous deepening of AI technology in the medical field, the correlation between the life and health of the patient and AI in public health institutions is becoming increasingly stronger. The confidential information data and major research achievements in research-based hospitals and scientific research institutions are increasingly relied on AI platforms. Ensuring the safety of patients and the transformation of research results by guaranteeing the stability of AI platforms will be an eternal proposition for these institutions and organizations.

4. **For AI R&D enterprises and biomedical business organizations, the fundamental need to promote AI empowerment in medical care is to improve the business efficiency of enterprises**

Whether it is an AI R&D enterprise or a biomedical business organization, the fundamental demand of all its codes of conduct is to ensure that the business of the enterprise conforms to the business logic and brings benefits to the employees and shareholders. Even if AI medical care can indeed bring great changes to the medical industry, greatly improve the status quo of the medical system and enable us to enjoy more intelligent, efficient and accurate medical and health services, will AI medical treatment be effectively promoted from top to bottom when the use of AI shows less attractiveness in profits for upstream enterprises and organizations of the AI industry? Taking the R&D of new drugs in biomedicine as an example, a long R&D cycle coupled with extremely tremendous cost and huge potential return are the typical features of venture capital investment.

5. **For society and public management, AI medical treatment may break the principle of fair distribution of social resources and reshape the current moral and ethical standards and cognition**

Overall, in the era of AI and big data, biomedical research is no longer limited to a few "experimental" samples, nor a few "exemplary" genes we are familiar with, but the first "lengthways synergy and the landscape orientation collaboration" with far-reaching significance in the history of biomedicine.

The so-called "lengthways" synergy refers to the number of individuals; that is, many individual diagnosis and treatment data, clinical data and "metadata" describing their living conditions are needed in the context of explaining the individual differences of diseases through digital technology to reveal the relationship between disease clinical manifestations and gene mutations and analyze the interaction between the occurrences of diseases and environmental variables. The so-called "landscape orientation" refers to the combination of multiple biomedical data of a single individual, such as a person's vital signs, personal medical history, physiological indicators, dynamic changes, etc. The "lengthways" synergy can determine the personalized characteristics of different patients, while the "landscape orientation" collaboration can explore the corresponding treatment methods. The combination and integration of the two can realize the integrated analyses of a wide variety of personal data of individuals and a single kind of data of many people. By doing so, more meaningful content can be excavated to form clinical diagnosis and treatment norms for the same disease as well as personalized disease prevention and treatment plans.

AI medical science is a specific practice of the "lengthways synergy and landscape orientation collaboration" of biomedical big data. Its realization requires the effective integration and sharing of multilevel and multityped biomedical data such as multiomics indicators such as health and disease files, molecular level as well as environmental factors to build a vast and dynamically evolving disease knowledge network and new taxonomy which can make patients with the same disease and the same biological basis benefit from a drug or other methods, and finally improve the health level and quality of the whole human society. Thus, in AI medical science, the sharing and use of individual patient data is indispensable. Only by publicly sharing data information can more welfare be brought to the experimental groups, their races and even the whole mankind. Of course, such data sharing needs to establish a perfect personal privacy protection mechanism that covers data anonymity and data desensitization of personal medical data.

It is worth mentioning that public management decision-makers have insufficient awareness of the value of AI ethics. Public management organizations and social governance decision-makers are generally aware of the urgent need for AI technology to energize traditional industries, but general attention has not been given to the AI ethics of various industries. Among them, technology ethics in the medical industry proves to be deeper, more multidimensional and comprehensive. For example, using AI to read and analyze gene data may create biochemical weapons that lead to gene variation. The products such as drugs or vaccines of biomedical experiments based solely on AI to simulate the biological organism environment without being experimented on human beings may involve the R&D of prohibited medicine. The analyses of medical data conducted by AI may involve information including race, gender and religious belief, which will result in group discrimination. Of course, there is anti-human relations and anti-human biomedical research and development.

Many human evils will continue to be amplified and cause serious consequences through AI. To effectively ease these problems, we need to let public management institutions and social governance decision-makers realize the profound value of AI ethics, actively understand and learn the content of ethics, constantly guide and intervene in all social parties involved in the application of AI medical science and strengthen the accountability and punishment for acts that challenge the basic social order and ethical bottom line.

Finally, let us summarize our thoughts on the emerging discipline of medical ethics. So-called medical ethics is the combination of medicine and ethics. Medical science has undergone the hardship of the primordial times, the clan era when doctors and witches can play the same role, and a long time of empirical medical science until today's sophisticated modern experimental medicine. During this process, the progress of medicine is always accompanied by medical ethics, and they have been even in perfect harmony. The great works of medicine are filled with a strong breadth of humanity. Whether the medical classic of ancient China, *The Medical Classic of the Yellow Emperor*, or the great work on internal medicine in modern western medicine, *Goldman's Cecil Medicine*, has dedicated a special portion to discussing medical ethics. Immortal doctors not only have superb medical expertise but also profound medical ethics. The spiritual banner of medicine is to cherish and revere life, and care is one of the main therapeutic methods in medicine.

Doctors and patients, a pair of moral strangers, need to adhere to their respective moral positions because rehabilitation is their common goal. Medical ethics is to see through all human medical activities with ethical eyes. Mankind has never enjoyed the benefits brought by medical science as it is today, and mankind has never faced such tough moral choices. Medicine is constantly changing the laws of nature by bringing dying people back to life. Medical science can turn a strong man into a graceful woman through countless surgical operations. Infertile individuals are able to have their own children with the help of medicine. All sorts of medical instruments can be used to help a person as fragile as grass continues his/her life for decades. We have lived for many years by transplanting other's heart and lungs. Scientists have come a long way, from research on the genetic chimerism between humans and animals to the cloning of human beings... "Science is a powerful instrument. How is it used, whether it is a blessing or a curse to mankind, depends on mankind and not on the instrument", Albert Einstein once said. Therefore, the medical ethics of AI also need to be considered from the perspective of medical ethics, constructing a research paradigm of medical ethics based on humanism.

4.3 Restart AI and Expand the Meaning Boundary of Life

After discussing the application and ethical issues of AI in medical treatment, we can see that AI has brilliant prospects in the medical industry. From a social perspective, the demand for medical treatment is hiking because of the aging population and the increase in chronic diseases. The prosperity of AI+ medical treatment can effectively ease the shortage of medical resources and improve the medical experience of patients. From the perspective of the medical industry, there are a large number of links in the traditional medical industry that need AI empowerment. The progress of A+ medical care can effectively drive the digital upgrading of the industry and improve the operation efficiency and achievement transformation of each link. From the view of the public, the development of AI+ medical care can effectively break the shackles of the conditions and resources of public health institutions on citizens' pursuit of a healthy life and realize all-weather and all regional health monitoring and personalized medical service experience. For a nation, a prosperous medical industry can effectively expand the dimension of industrial development, provide many new jobs and create new potential energy for the industry.

At the same time, the ethical issues of AI medical care, including privacy protection, physicians' dominant role in medical safety, platform stability, fair benefit and legal system construction, still need in-depth discussion and consensus in the industry. We need ethics to further propel the shaping of AI medical care, promoting technologies to serve a better life for mankind.

To optimize systematic engineering, deeply reconstruct the ethics system and effectively promote the implementation of standardization and oversight mechanisms, the first thing to do is to make all relevant parties inside and outside the industry fully contribute, exchange and study the principles of AI and basic knowledge information of the AI medical industry. Groups with professional abilities, such as AI R&D enterprises, universities and colleges, scientific research institutes, biomedical institutions and public health management and operation institutions, should give full play to their strengths, deeply share information and data from the following aspects: AI industry, academic research, bioscience and clinical diagnosis and treatment, medical ethics, public policy and social governance, elaborate knowledge achievements, and jointly discuss and build a targeted training and publicity system. Moreover, relevant parties should also learn from each other's strengths to update their knowledge. For example, hospital management departments can cooperate with AI enterprises to organize AI-related technology and ethics training seminars. AI enterprises can popularize AI technology principles and industrial development value to hospitals, and hospitals can communicate medical ethics and clinical diagnosis and treatment needs to AI enterprise personnel to enhance mutual understanding and interaction, thus reducing communication costs and improving cooperation efficiency.

Additionally, the government can also cooperate with hospitals, enterprises and other subjects, integrate publicity and education resources, and popularize AI technology and knowledge related to medical ethics to patients and the public through various activities, such as new media, short videos and popular science salons, to let patients, their relatives and the public have a clear understanding of the logic behind AI and the significance and value of AI in promoting the progress of the smart medical industry, alleviating the doubt, panic and resistance to medical treatment energized by AI due to lack of understanding, and making them understand, accept and apply AI medical treatment. In doing so, people can treat the great impact of science and technology on life and health more scientifically, rationally, objectively and dialectically and actively participate in the construction, development and supervision of AI medical ethics.

After all parties involved in the construction and development of AI medical industry have a comprehensive and full macro understanding and understanding of AI technology and medical health ethics, we need to cooperate with all parties to jointly establish a complete ethical norm system throughout the whole process of R&D and application of AI medical technology. Ethical norms should be incorporated into consideration from the approval of technical projects. The system must cover technical ethics risks, legal ethics, medical ethics, family ethics, public management ethics, social ethics, personal ethics, relevant evaluation standards and ethical crisis

4.3 Restart AI and Expand the Meaning Boundary of Life

response measures. Combining the technology research and development of AI with the ethical elements at many levels mentioned above, especially the standardized construction of medical ethics, will help to solve some ethical problems of AI medical technology and alleviate the concerns or excessive interpretation of these problems due to the lack of evaluation reference standards. The perfect ethical standard system also provides evidence for the legislative work related to AI, and the law enacted, in turn, can also make the key content in the standard system more practicable.

Technical and ethical advisory groups, industry associations and other institutions should also regularly evaluate the ethical risks and impacts on the medical industry empowered by AI to ensure the timely follow-up of relevant policies, laws, hospital management systems, ethical norms and technical standards. Meanwhile, public management organizations such as the drug administration and the National Centers for Disease Control and Prevention can jointly consider and formulate relevant AI and biomedical clinical medical technical specifications, safety standards, inspection and evaluation systems, risk prevention and control and crisis prevention countermeasures from multiple dimensions with standardized organizations, technical and ethical advisory groups, industry associations, medical institutions, enterprise representatives, experts, scholars, medical research institutions, universities, colleges, etc. Independent standardized organizations should clarify the application scope, scenarios and admittance threshold of AI technology, evaluate and check the product projects that do not meet the technical standards and ethical norms of AI medical care, dynamically monitor the ethical and technical risks of AI involving the medical industry, clarify the high-risk links, especially data encryption, strengthen the technical standards of data authorization management, and strongly encrypt and desensitize key privacy information in links such as training and result analyses. Optimizing the connection between service management and technical governance eliminates the potential safety hazards of AI medical treatment in an all-around way.

Technical safety standards and risk prevention and control strategies cannot play a constructive role in guiding the development of AI medical technology if they are not effectively implemented through independent and impartial supervisory institutions and high-cost punishment mechanisms. The quality standard supervisory organization can further improve the relevant supervision and consequence tracing mechanism according to ethical norms and technical standards. In the meantime, medical institutions, R&D enterprises and the whole industry should conduct self-inspection, self-correction, mutual supervision and promotion in R&D and the ethics application of AI and design the supervision system from the aspects of organizational structure, review system, management process, review norms and standards.

Let us look at the action measures at the legislative level. Traditional laws cannot regulate and restrict diverse problems derived from the AI medical industry. In view of the current situation that new business modes and demands are driven by technologies, the State Council has put forward the principle of prudent supervision, that is, appropriately and prudently relaxing policy and legal supervision in new fields, especially in the field of science and technology, to leave enough space for the growth of new technologies. As a link that has the most technical strength and application value in the new business, AI medical care needs enough space for development and breakthroughs at the legal level, rather than being firmly bound by the rules and regulations of law.

However, the prudential supervision here does not mean full liberalization. For acts of principle or those challenging the dignity of the law, the public security, procuratorial and judicial departments should implement necessary intervention measures in strict accordance with the relevant legal provisions. During the enactive, perfect and transitional periods of laws, relevant institutions should accurately and dynamically control the executive limits of laws, encourage bold technological innovation, and protect the legitimate rights and interests of the public, society and the state. Alternative measures targeting AI can be actively explored in laws, such as encouraging and recognizing the insurance system of AI medical products at the legal level, clarifying the rights and interests and compensation standards of insurance companies in case of medical accidents, and putting forward scientific, reasonable and fair exemption conditions and responsibility relationships targeting all parties involved in the application of AI in the medical industry, such as algorithm providers, manufacturers and enterprises responsible for production, to escort the vigorous progress of the AI medical industry.

As the core element of industrial development and quality technology foundation, standards play a fundamental and guiding role in the development of medical robots. Therefore, it is urgent to establish a set of medical robot standard systems based on an in-depth understanding of technical standards to provide a necessary basis for comprehensive, scientific and unified quality evaluation and to provide technical support for product listing approval. In addition, the system can effectively consolidate previous research achievements and improve the core competitiveness of China's medical robot industry by pushing its development toward a high-end orientation. Different from laws, the standard content here refers more to the ethical norm system and industry evaluation standards, mainly covering the technical maturity, the transparency of the algorithm model, product performance reliability, application effect, ethical risks and corresponding prevention and response measures of AI medical treatment, to lay a theoretical foundation for continuous improvements of legislation and supervision.

In short, to build a perfect top-level architecture standardization system of medical artificial intelligence, first, all parties in the industry need to fully understand, exchange opinions and reach a consensus on the definition, technical principles, needs, objectives, application scenarios, significance and value, potential risks and development route of the medical industry energized by AI. At this stage, because the medical and health industry energized by AI often faces the situation of interdisciplinary integration, the lack of a top-level design of forward-looking ethical research and ethical governance, and the direct relationship between the medical industry and citizen life security and social welfare, we tend to be conservative in the face of digital intelligent transformation. Finally, the long-term absence of ethics systems and industry evaluation standards may restrict the development of artificial intelligence legislation and artificial intelligence supervision systems in the medical industry.

Looking ahead, the continuous changes in the global energy mix will make the future computing power eliminate energy limits. The R&D of new materials and the realization of quantum computing will not only make AI break through the limitations of carriers made of semiconductive materials and the logical constraints of data structure but also drive the development of human science and technology into a new era, exerting subversive impacts on the development of human society, psychological construction and cognitive breakthroughs. At the moment, we cannot help thinking: the ultimate meaning of the technical revolution is to meet the evolutionary needs of human beings to better adapt to the environment, enjoy a higher standard of living, and bring benefits to human beings. Or is it a technology and capital bubble created and driven by interests to highlight the technological superiority of individuals or organizations? Are the multiple industry applications of health care energized by AI accurate solutions to the pain spot of the industry or the pseudo proposition and demand of simply pursuing technological innovation and capital expansion?

Taking the gene editing baby event in 2008 as an example, He Jiankui, an associate professor of Southern University of Science and Technology, conducted technical research on AIDS prevention by editing human embryo genes. Although some scholars held reservations about this event, they stood out and criticized that it was a serious violation of medical ethics and challenged the bottom line of moral values for pursuing personal fame and academic achievements. This incident warned that AI's deep participation in biomedical construction should be kept in certain ethical limits. Science and technology should adhere to people-oriented development intentions instead of overriding human morality and ethics and alienating human beings into tools or even experimental products. From the point of human survival and progress, the application of AI in the medical field equals a profound social moral experiment. In view of the potential medical moral and legal risks in AI applications, the international community has put forward a series of schemes to try to control intelligent machines on the issue of "how to create intelligent moral machines" and gradually reached a consensus. Human moral values and ethical norms can be "embedded" into intelligent machines with the help of intelligent algorithms to let them have human-like ethics and morality such as morality, shame, responsibility and compassion.

To achieve this goal, it is urgent to enable intelligent machines to reason morally in real AI medical application scenes by embedding the values, moral codes and ethics obtained by human beings in medical scenarios through algorithms.

These above are some of our thoughts on this field. In fact, there are many AI- and medical-related fields we have not discussed here, such as the relationship between artificial biology and artificial intelligence, which is a more cutting-edge field that determines our future and an important field for us to understand the new paradigm of AI and man-machine relationships. The chapter aims to help you learn about the risks and challenges faced by human beings in AI application in the medical field and trigger the thought on how we should face these challenges. It is believed that with the continuous application of AI technology in the medical field and its deep integration with other related technologies, we can expect a more orderly and loving (AI) society and a healthier and more meaningful life experience.

Chapter 5
"Secret" Left by Turing—Privacy Computing

AI has been applied in an increasing number of fields and scenarios, such as automatic driving, medical treatment, media, finance, industrial robots, and Internet services. On the one hand, it has improved efficiency and cut costs; on the other hand, the autonomy of AI systems makes algorithm decision-making gradually replace human decision-making, which sometimes leaves the existing problems undone or even harder to be dealt with, and completely new problems may also emerge in society. All these old and new problems not only cause extensive discussion in society but also become important factors restricting the implementation of AI technology. In the face of AI technology with as great risks as its potential, people urgently need a broad and universal ethical discussion to find paths and comb the norms to guarantee the benign development of AI.

In recent years, the "materialization of morality", a frontier hot topic in the international philosophy of technology, has attracted considerable attention at home and abroad. The core of the concept is practicing morality through the use, layout and popularity of technical artifacts embedded with specific value through design. Undoubtedly, the best practice object of this idea lies in "AI" technology. Traditional negative ethics is transformed into positive ethics by introducing ethical considerations and keeping the black box open at the design stage. This is not only the basis for us to study the ethical problems of AI in this chapter but also the difference from the previous traditional ethical propositions.

Based on the above propositions, this chapter will discuss how deep learning technology develops further in the direction of respecting people's values and rights from the perspective of ideas under the dual regulation of data ethics and algorithm ethics. AI technology has gradually formed an advanced privacy protection path, such as privacy computing. In the technical paradigm, with the support of privacy-enhancing technology, a value leap can be realized while following the principle of putting people first.

© The Author(s), under exclusive license to Springer Nature Singapore Pte Ltd. 2022
Z. Liu and Y. Zheng, *AI Ethics and Governance*,
https://doi.org/10.1007/978-981-19-2531-3_5

5.1 Ethical Turn of In-depth Learning: Theory of Materialization of Morality

With the deepening of research on AI, the research field of AI is also expanding. Throughout history, many branches of AI research have emerged, including recommendation systems, machine learning, evolutionary computing, fuzzy logic, computer vision, and natural language processing. However, all the current scientific research work is focusing on weak AI, which cannot be regarded as equal to the strong AI with independent thinking and emotion in movies. Weak AI has broken through and become the mainstream of current AI, and much of the credit goes to machine learning, an approach to realizing AI.

Although the applications of traditional machine learning algorithms in fingerprint recognition, face detection, object detection and other fields have basically reached the commercialization requirements or the commercialization level of specific scenes, every step forward proves to be difficult until the emergence of a deep learning algorithm, which is a neural network used to establish and simulate the analyses and learning functions of the human brain, and a machine learning technique that simulates the mechanism of the human brain to interpret data.

The basic feature of deep learning is that it is based on the representation of data to imitate the transmission and processing of information between neurons in the human brain. The most significant applications are computer vision and natural language processing (NLP). Deep learning is divided into convolutional neural networks (CNNs) and deep belief nets (DBNs). The main idea is to simulate human neurons. Each neuron receives information and transmits it to all its neighbors after processing. Deep learning uses the idea of hierarchical abstraction that higher-level concepts learn from lower-level ones. A hierarchical structure is often constructed by greedy algorithms layer by layer, and key point capture is conducive to the efficiency of machine learning. Since most deep learning exists in the form of unsupervised learning, they have unique advantages in dealing with unstructured and unlabeled data.

Thus, we can see that among the basic production factors of deep learning, the two factors highly related to users are data and algorithm. In addition, the deep learning algorithm directly determines the impact of the model on the main subject it acts on. The premise of deep learning algorithm ethics is that the value of the algorithm is not neutral but value loaded. Algorithms can trigger new technical ethics problems or aggravate the original ethics problems as well as eliminate or resolve them.

Specifically, the related ethical problems mainly have the following three forms: presupposing a certain value position in the algorithm, the running result of an algorithm has some ethical effects (such as information cocoons, conformity) and algorithms constructed in a social order independent of the governance of a real country. The manifestations above will cause problems such as improper behavior, opacity, discrimination and prejudice, information privacy challenges, dilution of moral responsibility and so on.

5.1 Ethical Turn of In-depth Learning: Theory of Materialization of Morality

To correctly guide the ethical trend of AI relying on deep learning, we can understand it through the theory of "materialization of morality". Let us first look at the difference between the theory of materialization of morality and the traditional ethics of science and technology.

The traditional ethics of science and technology adopt the idea of technology evaluation, which uses ethics as a review standard to evaluate its severity. It is a posterior governance strategy. As cutting-edge technologies are breaking the existing legal and ethical frameworks, such remedial solutions afterwards cannot predict the possible ethical dilemmas caused by technology in advance. The theory of materialization of morality provides a prior solution. More specifically, it provides a positive ethical idea that can "regulate" mankind through the design of technical artifacts to make people become more moral by the role of technology and realize the ultimate goal of human-omputer interaction.

From the idea of the philosophy of technology, it breaks the traditional perspective of binary opposition of the subject and employs the perspective of regulation to dissect the ethics of technological material desire, which represents its innovation. The relevant theories have been deeply discussed among the technical philosophers of the Dutch school. For example, there is a very in-depth discussion in the book *"Moralizing Technology: Understanding and Designing the Morality of Things"* written by a famous philosopher, Peter-Paul Verbeek.

In fact, Hans Achterhuis, the teacher of Verbeek, created the concept of the "materialization of morality". His famous philosophical work *"American Philosophy of Technology: The Empirical Turn"* is a topic discussing the transformation of the philosophy of technology from traditional grand narrative and technical metaphysics to empirical research. He first mentioned "materialization of morality" in his discussion on the relationship between climate impact and carbon emissions in the Netherlands, a low-lying country, in his article in 1995. Given our current concern about "carbon emission" and "carbon neutrality", it is clear that the thinking of the technological philosopher is at the forefront. What he focuses on is to create a positive material environment through the design of technical artifacts and social systems to make it easier to practice behaviors that meet moral requirements. This is also the basic idea for us to discuss the solution to AI ethical risks.

Currently, ethics should gradually shift from theoretical reasoning to practical application, especially how to transcend the specific technical state and form ethical judgments and decisions with high execution from the angle of design based on practical moral problems. If the goal is to promote the results of applied ethical analysis to be implemented in concrete and practical propositions, what institutional and practical material conditions are needed, and how to design systems, components, infrastructure and applications to keep an ethical concept stable in long-term reflection and evolution in the future?

The above ethical transformation that adapts to engineering propositions in reality makes applied ethics need to consider not only applied analysis but also economic conditions, technology, systems, legal frameworks and social governance codes, especially the design of technical products and sociotechnical systems. Meanwhile, with the rise of information technology, digital technologies represented by big data and AI technology have been closely related to design. The value dimension of information technology, equal to the value and real needs of society and users in the application of Internet technology to production activities, has become a significant indicator in evaluating technical feasibility.

Therefore, how can value and moral dimension design be applied to AI and big data technology? Based on the method system combining concept, experience and technology, the author mainly promotes it from the following three aspects.

From the conceptual level, the concept of AI ethics is the main subject of analysis. Before the implementation of AI technology, what should be clarified are the direct and indirect stakeholders affected by the design of the technology and how the designer can weigh the conflicts and needs of these stakeholders in autonomy, security, privacy and other rights in the design, implementation and application of AI technology. Second, we should think about how to avoid risk by optimizing design. For example, driving data collected by users when we develop autopilot driving can help optimize driving experience and road safety. Is such data transferation feasible at the value level? Should these driving data be destroyed in time after being used? Avoiding privacy leakage, improving transparency, avoiding discrimination and unfair distribution are all value-sensitive concepts that should be considered in the initial stage of technology and ethics research.

From the empirical level, the practice of AI ethics is the main subject of analysis. First, it is necessary to conduct a practical analysis of the social environment in which technical products exist, which means evaluating the success and failure of a specific design by observing, measuring and recording the activities under constraints. Then, we need to fully implement ethics and responsibilities from the development, promotion, strategy, policy, personnel organization and other aspects of AI technology. AI technology should be developed in accordance with ethical principles, and responsible innovation will effectively ensure the interests of both individuals and groups. How to make trade-offs and whether to give priority to individual value or feasibility in the face of competing or conflicting value designs. Fully respecting the core value and interest appeal of each class group and designing a certain incentive mechanism to boost the whole system are of great significance.

From the technical level, algorithm ethics is the main subject of analysis. We should analyze whether a given technology is biased toward a certain moral value and how to support or hinder a particular value according to the value adaptability provided by technical characteristics. The increasingly complex algorithm of AI brings about increasingly diversified ethical problems, which puts forward higher ethical and moral requirements for algorithm design.

The development and application of AI technology are profoundly changing human life, which will inevitably impact the existing ethics and social order and trigger a series of problems. Among them, there are not only intuitive short-term risks, such as security risks caused by algorithm vulnerabilities and discriminatory policy formulation arising from algorithm bias but also relatively indirect long-term risks, such as the impacts on property rights, competition, employment and even

5.1 Ethical Turn of In-depth Learning: Theory of Materialization of Morality

social structure. Although short-term risks are more concrete and sensible, the social impact of long-term risks turns out to be more extensive and far-reaching and should also be taken seriously.

In the long run, the ethical risks of AI application are unique. First, it is closely tied to personal interests. For example, if algorithms are applied to crime evaluation, credit loan, employment evaluation and other occasions concerned about personal interests, personal rights and interests will be systematically damaged once discrimination occurs. Second, the causes of algorithm discrimination are usually difficult to determine. Deep learning is a typical "black box" algorithm. Even the designer may not know how an algorithm makes decisions. It is technically difficult to identify whether there is discrimination and its sources in the system. Third, as AI is increasingly widely being used in enterprise decision-making, the profit-seeking nature of capital is more likely to infringe on public rights and interests. For instance, phenomena such as price discrimination against customers based on user behavior data analyses conducted by enterprises and targeting users with ads for games, addictions and even fake dating sites through AI to gain huge profits may occur.

Next, we will discuss how to employ the thinking of materialization of morality to solve ethical problems in deep learning. Actually, we have benefited a lot from deep learning technology, but we should also pay attention to its potential ethical, moral and legal risks and challenges above.

These risks and challenges can be classified and summarized into the following six aspects:

Data security risks include the risks of data inside the algorithm model caused by reverse attack;
Network security risks, including the risks of network security vulnerabilities in the AI learning framework and components;
Algorithm safety risks—rong design or implementation of an algorithm can produce inconsistent and even harmful results;
Information security risks—AI technology may be used to make false information and commit fraud;
Social security risks—AI may set off serious social moral problems and endanger social security;
National security risks—AI may be used to manipulate public opinion and indirectly threaten national security.

On the one hand, with the infiltration of AI, there are still no clear answers to a series of events that have happened, such as the definition of medical accident responsibility of the "Leonardo Da Vinci" surgical robot, the ownership of the copyright of the "XiaoIce" poetry collection, and the ownership and responsibility of human death caused by Uber unmanned motor vehicles. To avoid the moral hazards of AI, the idea of a moral machine has begun to emerge and received much support, but the view is based on emotion rather than rationality. The proposal of machine morality stems from human concern and panic about the development of AI technology. We hope that machines can be so moral that they will not harm us while serving us.

Machine morality is the expansion of human morality in the era of AI, while artificial intelligence is artifactitious. Whether AI has human sociality is still controversial in the academic community. Some scholars believe that AI under deep learning will possess independent moral status with "free will" in the future. Some scholars hope to realize a "moral machine" by embedding a moral algorithm in a deep learning algorithm, which conforms to the idea of materialization of morality.

We need to build a set of highly executable machine ethics mechanisms to allow machines to make ethical behaviors by themselves. Since general or strong AI has not been realized at present, we need to transform morality and responsibility into a behavior standard and specification that can be operated and executed by machine with the help of the machine codes of data and logic. The process is something like converting ethical theories or norms advocated or accepted by human beings into ethical algorithms and operation processes that can be calculated and executed by machines, then describing multiple values and theoretical categories of computers by quantity, logic, probability, etc., and finally writing ethical algorithms and embedding them into machines.

From the practical level, the embeddedness of machine ethics mainly has the following three points: first, preset a set of sufficiently operational ethical norms, and then use machine learning methods such as reinforcement learning to let AI machines study human-related reality and simulate behaviors in various scenarios and establish values and adopt strategies that are similar to human beings. Finally, build an effective human-computer interaction mechanism to enable machines to fully explain the rationality of their decision-making. In the meantime, human beings can correct or intervene in time when necessary to avoid the loss of control of machine independent decision-making in complicated scenes.

At the same time, embeddedness embodies not only the machine itself but also the leading role of human beings in the whole process, including the development, production, sales, use and destruction of machines. In addition, a number of links, such as ethical algorithms, ethical evaluation and testing tools, human-computer interaction interfaces and ethical supervision, should all be incorporated into the embeddedness.

In reality, since AI was first proposed in the 1950s, it still remains in the weak stage after half a century of development. Technologically, the next generation of AI is likely to evolve in the following three directions.

In terms of the underlying technology paradigm, the first is AI technology that can realize complex computing based on dynamic environment changes, including big data intelligence, swarm intelligence, cross media intelligence, hybrid-augmented intelligence and intelligent unmanned systems, which are also the new generation of AI technology being promoted in China.

In terms of the application target scenario, the second is AI technology based on new application paradigms formed under complex goals that are aimed at urban groups. At present, typical examples are the "Woven City" pushed by Toyota in Japan and the future city "Neom" accelerated by Saudi Arabia.

In terms of basic chip architecture, the third is revolutionary AI technology based on the new chip architecture. In the Von Neumann architecture, the mainstream architecture of AI chips, there is a ceiling of memory performance due to separated units of computing and memory. The result also leads to limited growth space for processor performance and affects the effectiveness of the computing power of the current neural network model for complex scenario processing. Thus, the adoption of in-memory computing in the new chip architecture is of great value.

Finally, we summarize the value and risks of the technical philosophy of materialization of morality to the practice of AI ethics and governance. On the positive side, we can see that this theory is different from the traditional grand narrative philosophy in its operability. The theoretical thinking of the concept is more or less affected by other relevant theories, such as Batya Friedman's value-sensitive design, B.J. Fogg's persuasive technology theory, Wiebe Bijker's social construction of technology theory (SCOT) and so on. All these theories have recognized that technology is not a value neutral tool but is highly interconnected with the social value system. Of course, the theory also has its risks and limitations. Once technical artifacts are written into the logic of value preference and form a consensus, it is likely to lead to value industrialization or centralization. In that case, engineers will be given greater rights. Additionally, due to the restriction of technical barriers, it is obvious that the general public knows little about its consequences, thus forming a black box for technology, which may also bring about two risks. One is that the technology loaded with specific value may be dominated by the elite, resulting in the concentration of value rights. The other seems more dangerous. Value propositions may be controlled by sovereign states, leading to the birth of value colonialism.

5.2 Fantasia in the Turing Era—From Encryption to Computing

The Anglo-American film *The Limitation Game* released in 2014 tells a true story about British mathematician Alan Turing, who helped design an electronic computer 60 years ago to crack the military code of Nazi Germany during World War II. The title of the film has nothing to do with Turing's achievements in the film directly but comes from a popular game in Britain at that time. In the game, a man and a woman hide behind a curtain, and other game participants keep asking them questions. The two hidden need to answer these questions in illegible handwriting. After that, let questioners try to figure out their gender according to the answers.

In 1950, Turing borrowed the form of this game as the standard to judge whether a computer has human intelligence in his paper *Computing Machinery and Intelligence*. He placed a person and a computer behind the curtain and let the tester judge which one was a computer by asking questions. If the judgment was wrong, the computer was regarded to have passed the Turing test and possessed human intelligence. AI scholars later called the computer described in Turing's paper Turing machine and this kind of test Turing test.

However, the Turing test set a goal for the development of AI. With the continuous research of several generations of AI scholars, people have gradually realized the high complexity of the human brain and the limitations of computers. These findings help us continue to apply AI technology to many aspects of production and life. This section will help you understand Turing's impact on AI technology and his ideas and solutions in privacy encryption by reviewing and discussing his research.

In 1939, the First World War broke out, and the world was trapped in an anxious war. The information war played an important part in the Second World War, so the encryption and decryption of war information was crucial and even determined the victory or defeat of the war. By then, Nazis took advantage of their superior advantages in information dissemination. This was entirely because they had the second generation of Enigma encryption password, which was the main communication encryption tool assigned to all military departments in Germany. All military information of Nazi German troops was transmitted through this kind of password.

The encryptor encrypts and transmits information through different combinations of its tangents and rotors; that is, it can easily realize different encryption logics by changing the wiring and rotors. The German army kept changing encryption logic on time every day through a diversity of combinations. Therefore, it was difficult for cryptologists in other countries to decipher their passwords. When countries were ignorant of the military intelligence spread by the Nazis, a genius stepped forward bravely, solved the problem and ended the Second World War as soon as possible, saving a myriad of lives.

He is Alan Matheson Turing, a famous mathematician and logician born in England. In 1939, Britain was in the fog of war. Due to the needs of the war, Britain established a government password school specializing in deciphering the Nazi German military password system. The school was composed of the intelligence departments of the British army and the Royal Navy by then. As a famous mathematician in his era, Turing was naturally recruited by the British government to join the government password school. Due to the complexity of cipher encryption methods and the timeliness of intelligence, Turing and his comrades in arms theoretically only have 24 h to crack each password. Even if the password was finally cracked, some military intelligence lost value due to timeliness.

At one time, code decoding reached an impasse. The government password school gathered a large number of top mathematicians and cryptologists in Britain for code decoding work. They sat together every day to brainstorm and discuss in the hope of working out feasible schemes for code decoding. However, Turing behaved exceptionally eccentrically at that time. He never had in-depth communication with others on any problem inside or outside working hours. In addition, as Turing was allergic to pollen, he wore a gas mask when cycled to work every day. All these characters made him become a "monster" in the eyes of others. It is this "monster" that was making a decoding machine named "Bombe" that could replace manpower and speed up the decoding efficiency.

It was a decoding machine made up of 36 enigma that could run simultaneously. The machine could run its rotors at a high speed and examine possibilities one by one to find password patterns. While the machine greatly increased decoding speed,

5.2 Fantasia in the Turing Era—From Encryption to Computing

Turing was still not satisfied with the result. To break through the second-generation enigma, he had to work out a machine that had higher computing speed.

Therefore, he focused on the secret telegrams that had been decoded and tried to find a breakthrough by them. Through continuous analyses, Turing finally found the words repeatedly used by the Nazi Germans in passwords, including the words about weather and peace. He found that "hrer" was mentioned in almost every coded information. Therefore, Turing realized that by corresponding these words appears almost every day in their telegrams and deciphering the encryption law of the single word, the encryption logic of all daily information in Nazi Germany can be accessed. After trial and error, the method considerably reduced the amount of computation required for decoding. As a result, it took only one hour at most to crack all the military secrets of the Nazi German army.

In August 1940, Turing's "Bombe" decoder finally displayed its power to the full. He also cracked the password system of the Nazi German military with his genius. Due to the particularity of this task, his achievements were kept as secrets in case the German army noticed.

In 1941, the German military cryptosystem changed suddenly. Since the whole coding logic became more complex, the decoder became much less efficient than before. Turing suddenly aware that the encryption combination of German second-generation passwords had changed, so a more efficient decoder must be put into use. Considering all this, Turing found his teacher Max Newman to jointly develop and produce a more effective decoding machine. Soon, they made a supercomputer named "Sith Robinson", which was faster than "Bombe", and accelerated its computing speed with the help of engineer Thomas Harold Flowers. It is because the machine had deciphered Hitler's secret information before the end of World War II that the Allies could successfully bypass the heavily fortified Galle Beach in Germany and chose to land in the least defended Normandy to launch the full-scale counterattack against the Nazi German army that had already successfully landed there, ending the main war on the European continent the following year.

In fact, the idea of creating a machine that looks like a human or mythical figure with powers far beyond humans has been around since ancient times. It was only after the emergence of the material basis for building machines with "thinking" ability that the idea gradually came to fruition. Turing took the lead in answering what kind of machine is capable of "thinking". In his opinion, a machine capable of performing logical operations can be a "thinking" machine. He believed that the essence or core of human thinking is logical operation. As one of the pioneers of AI and modern cryptography, Turing actually implied the relationship between encryption and AI. Turing's definition of an intelligent machine is based on the implicit assumption that human thinking is a quantifiable structure. In his paper, instead of answering what "thinking" is directly, Turing explained it in an experimental way bypassing endless questions of philosophers. In short, he thought that a machine that could pass the Turing test was a machine that could "think".

In his paper published in 1950, Turing pointed out that readers must accept the fact that digital computers can be constructed and that this has already happened. According to the principles we describe, they can actually simulate human computing behavior in a very similar way. We note that Turing's intelligent machine must do at least three things to play its role: first, researchers should quantitatively deconstruct human thoughts and give out corresponding mathematical formulas; second, programmers translate these formulas into a series of commands that can be executed by computer; finally, these commands can be stored, computed and executed by computer.

After Alphago defeated Lee Se-dol, a top human go player in 2016, we witnessed the great potential of AI and began to look forward to more complicated and cutting-edge AI technology applications in many fields, including pilotless automobiles, medical treatment, finance, etc. However, today's AI still faces two major challenges. One is isolated data islands in most industries, and the other is data privacy and security. Personal social media information, medical and health information, financial information, location information, biometric information, consumer portrait information, etc., are often overshared and abused in the era of the digital economy, and enterprises or institutions collecting and processing this information often lack sufficient privacy encryption and protection capabilities. Meanwhile, as the global awareness of data value increases, data privacy and security have become an important cornerstone of enterprise business operations, and their significance cannot be overemphasized. For this reason, with the increasingly strong compromise consciousness of large companies in data security and user privacy, attaching importance to them has become a major problem worldwide.

Information about public data leakage has aroused great concern from the public media and the government. A survey conducted by the Pew Research Center last year found that 79% of adults were worried about how companies use the data collected about them, and 52% said they give up using some products or services for the concerns about their personal information being collected. The traditional data processing model in AI often involves a simple data transaction model. One party collects data and transmits it to the other party responsible for data cleaning and fusion. The third party will obtain the integrated data and build models for other parties to use. These models are usually sold as final service products. This traditional process faces challenges from new data regulations and laws. Moreover, since users may be unclear about the future use of the models, these transactions violate laws such as the EU's *GDPR*.

At present, we are faced with a dilemma: our data exist as isolated islands, but as global data compliance regulations become stricter, we are prohibited from collecting, integrating and using data for AI processing in many cases. Moreover, there are also frequent privacy disclosure events that deepen the trust gap. Legally solving the problems of data fragmentation and data islands is the main challenge faced by current AI researchers and practitioners.

5.3 Secret Greatness: Creating Privacy Enhancing Technology

How to break the existing data barriers upstream and downstream of the industrial chain, effectively solve the problems of competition and monopoly in the digital market, fully stimulate the value of data elements, share digital dividends, and realize "Land to the Tiller" in the era of the digital economy have become the focus of all sectors of society. The increasingly strict supervision of privacy protection has not only promoted the awakening of privacy protection awareness of data rights subjects and data processing behavior organizers but also exacerbated the concerns of enterprises on the legal compliance of data circulation and cooperation. In this context, privacy computing comes into being. It can complete the computing and analysis tasks based on data and information while ensuring privacy security.

As privacy issues are gradually standing out, relevant legislation is also progressing steadily in response to privacy risks. On May 25, 2018, the EU issued the *GDPR,* which stipulates users' right to access data, right to be forgotten, right to restrict processing and right to migrate data to protect personal privacy and curb data abuse. On April 16, 2019, San Francisco passed some amendments to the *Stop Secret Surveillance Ordinance.* Given that face recognition technology may infringe on users' privacy and aggravate racial discrimination, it is prohibited in the city. The State Internet Information Office of China issued the *Measures for Data Security Management (Draft for Comments)*, discussing measures from the aspects of data collection, processing and use and security supervision on May 28, 2019. Therefore, privacy protection technology represented by privacy computing is becoming increasingly active.

After decades of development, privacy computing is playing an important part in industrial Internet, AI, financial technology, medical protection, shared data, etc. It can be divided into federated learning, secure multiparty computation, confidential computing, differential privacy technology, homomorphic encryption technology and so on.

Federated learning was first proposed by Google in 2016. It was originally used for Android mobile phone users to update models locally. Federated learning means that a central server coordinates multiple loosely structured intelligent terminals to update the language prediction model. Its working principle is as follows: the client downloads the existing prediction model from the central server, trains the model with local data, and uploads the updates of the model to the cloud. The training model optimizes the prediction model by integrating the model updates from different terminals, and then the client downloads the updated model locally. The process repeats itself. As endpoint data are always stored locally throughout the process, there is no risk of data leakage.

Federated learning is essentially a distributed machine learning technology or machine learning framework based on the principle of minimum data collection. It involves training statistical models through remote devices or isolated data centers (such as mobile phones or hospitals) while maintaining the localization of data. Federated learning can generally be understood as a technical architecture that involves two or more participants to cooperate to build and use machine learning models while ensuring that the original data from different parties does not exceed its defined security control scope. It is a special distributed machine learning architecture that not only maintains the decentralized distribution of training data but also guarantees the data privacy of participants. Compared with the centralized training model, the machine learning model based on federated learning has almost lossless performance. The goal of federated learning is to realize the collaborative modeling of AI models and improve its effectiveness on the basis of ensuring data privacy security and legal compliance.

Secure multiparty computation (MPC) was first officially proposed by Yao Zhi, a Turing Award winner and academician of the Chinese Academy of Sciences, in 1982 to solve the problem of privacy protection and collaborative computing among a group of participants distrustful of each other. During the process, independent input, correct and safe computation must be guaranteed. MPC is mainly used to safely compute conventional functions without a trusted third party. At the same time, it requires that each participant cannot obtain any input information from entities other than the calculated result. MPC plays an important role in multiple scenarios, such as electronic election, electronic voting, electronic auction, secret sharing and threshold signature.

Specifically, each MPC participating node in the same status can initiate collaborative computing tasks or choose to participate in computing tasks initiated by other parties. Routing addressing and computing logic transmission are controlled by hub nodes that transmit computing logic while looking for relevant data. Each MPC node completes data extraction and computation in a local database according to computing logic and routes the output computed result to a specified node so that multiparty nodes can complete the collaborative computing task and output unique results. In the whole process, the data of all parties are local and will not be provided to other nodes. On the premise of ensuring data privacy, the computed result is fed back to the whole computing task system to allow all parties to receive correct data feedback.

Another technology that deserves attention is confidential computing. In the past, security measures mainly focused on protecting static or encrypted data for transmission. In fact, encrypting data in databases and data transmitted over LAN/WANs and 5G networks are key components of all such systems. Almost all computing systems (even smartphones) have built-in data encryption functions, which are enhanced through a special computing engine in the processor chip. However, if malicious users access device hardware or invade a device without using an encryption function (a relatively neglected area) through malicious applications, then all these encryption functions will be out of action. If machine memory can be accessed at this point, all data can be easily viewed and copied.

The original intention of confidential computing (CC) is to eliminate this potential risk. In August 2019, Linux Foundation announced the establishment of Confidential Computing Consortium (CCC) composed of a number of technology giants

5.3 Secret Greatness: Creating Privacy Enhancing Technology

including Accenture, Ango, ARM, Google, Facebook, Huawei, Microsoft and Red Hat. The organization is committed to protecting the security of data applications in cloud services and hardware ecosystems. Before defining the CCC, some organizations have defined confidential computing. For example, Gartner defined confidential computing in its *Hype Cycle for Privacy, 2019* as follows: "confidential computing is a combination of CPU-based hardware technology, IaaS cloud service provider virtual machine images and related software which enable cloud service consumers to successfully create an independent and trusted execution environment, also known as enclave. By providing a form of encryption in data use, these enclaves can make sensitive information invisible to host operating systems and cloud providers". The trusted computing alliance believes that confidential computing should cover a wider range of application scenarios other than cloud computing. Additionally, the use of the term "encryption" is not that rigorous because "encryption" is only one of the technologies for data privacy protection but not the only technology. The technologies of confidential computing should include other technologies being explored.

Therefore, the CCC defines confidential computing as "one of the technologies to protect the privacy of data applications by performing computing in a hardware-based trusted execution environment". To reduce the trust dependence of the confidential computing environment on proprietary software, the research of confidential computing focuses on the security assurance of hardware-based executable environments. The hardware-based trusted execution environment (TEE) is the core technology of confidential computing. It provides an isolated execution environment based on hardware protection, which has gradually become the focus of attention in recent years. According to industry practice, TEE is defined by the CCC as an environment that provides a certain level of protection in terms of data confidentiality, data and code integrity. Some more mature technologies that have introduced TEE include ARM's TrustZone and Intel's software guard extensions (SGX).

The next technology worth discussing is "differential privacy technology". To deal with user privacy disclosure brought by the increasingly advanced information society, the differential privacy model, a widely recognized strict privacy protection model, protects the potential user privacy information in the published data by adding interfering noise. Thus, even if the attacker has information other than specific information, he/she still cannot infer the information. Hence, this is a method that can completely eliminate the possibility of the disclosure of private information from data sources.

Differential privacy is a privacy definition based on strict mathematical theory. It was designed to ensure that attackers cannot infer any sensitive information related to an individual according to output differences. That is, differential privacy must provide statistical indistinguishability of output results. Any differential privacy protection algorithm is inseparable from randomness, so no deterministic algorithm can realize the indistinguishability of differential privacy protection. Differential privacy only realizes privacy protection by adding noise without additional computational overhead, but it still has a certain impact on the availability of model data. How to design a scheme that can better balance privacy and availability is also the center of attention in the future.

Differential privacy technology is usually used to handle the privacy protection of individual queries. However, in practical applications, it is often necessary to merge multiple privacy computations on the same data set or repeatedly execute the same privacy computation and determine what degree of privacy protection can be achieved. The proposed resultant theorem aims to combine a series of computations that meet the differential privacy requirements while ensuring that the differential privacy requirements are met in general. Most traditional differential privacy schemes are centralized differential privacy schemes, which means noise is usually added to data by a trusted third party. To reduce the demand for trusted third parties in practical applications, some decentralized privacy protection schemes, such as local differential privacy, have been put forward in recent years.

The last technology we discuss is "homomorphic encryption technology". Homomorphic encryption, a special encryption method, allows the ciphertext to be processed to obtain a result that is still encrypted. That is, processing ciphertext directly yields the same result as processing and encrypting plaintext. Homomorphism is maintained from the perspective of abstract algebra. The concept of homomorphic encryption is a cryptographic technology based on the computational complexity theory of mathematical puzzles. It can be simply explained as follows: ciphertext of homomorphic encrypted data is processed to obtain an output decrypted in the same way as the output obtained from the unencrypted raw data processed in the same way. At present, homomorphic encryption is mostly realized by asymmetric encryption algorithms. In other words, all participants who know the public key can encrypt and perform ciphertext computation, but only the private key owner can decrypt. According to the supported functions, the current homomorphic encryption schemes can be divided into somewhat homomorphic encryption (SWHE) and fully homomorphic encryption (FHE). For computer operations, achieving complete homomorphism means that all processes can achieve homomorphism. Homogeneity that can only be realized in some specific operations is called somewhat homomorphism. Algorithms with somewhat homomorphic properties include RSA, Elgamal, Paillier, Pedersen commitment, etc.

Homomorphic encryption technology was first used in cloud computing and big data. Homomorphic encryption is also a good supplement to blockchain technology. With homomorphic encryption, the smart contract running on the blockchain can process ciphertext without accessing real data, which greatly improves privacy security.

Generally, privacy-enhancing computation has become a standard concern of academia and industry in recent years. Privacy-enhancing computation completes the computations and analyses of data on the premise of data privacy protection through system security technologies and cryptography technologies such as federated learning, secure multiparty computation, confidential computing, differential privacy technology, homomorphic encryption technology and so on. Enterprises are not only in the 2C market directly facing consumers but also looking for ways to reduce privacy risks and concerns in the B2B environment, which stimulates rapid progress and commercialization in privacy-enhancing computation (PEC). As a powerful technology category, PEC can be used to enable, enhance and protect data

5.3 Secret Greatness: Creating Privacy Enhancing Technology

privacy throughout the life cycle of products and services of enterprises. By adopting data-centric privacy and security methods, these technologies contribute to ensuring that sensitive data are effectively protected during processing.

PEC not only includes system security technology and cryptography technology but also information collection, storage, and data security technologies protecting and enhancing privacy security during search or analyses. Many of these technologies have overlaps or can be used in combination. Although there are some differences in the security of privacy-enhancing technology in various applications and use cases, generally speaking, the more secure the technology is, the more privacy protection or privacy protection functions it provides.

The content above is our basic introduction and discussion of privacy computing technology. Simply put, privacy computing realizes the "availability and invisibility" of data at the same time through technology so that data from different sources can be safely shared to produce greater value, including MPC based on cryptography and the single or comprehensive use of various technologies such as federated learning derived from AI. The current general consensus in the industry is that to realize the "availability and invisibility" of data, it is difficult for a single technology to provide all the support needed, and the complementary integration of different technology paths (cryptography, artificial intelligence, blockchain, etc.) should be the development trend.

Most of the existing privacy protection schemes center on relatively isolated application scenarios and technical points and propose solutions to specific problems existing in a given application scenario. The privacy protection scheme based on access control technology is suitable for a single information system, and privacy protection problems in metadata storage and dissemination have not yet been solved. The privacy protection scheme based on cryptography is also only applicable to a single information system. Although with the help of the key management provided by a trusted third party, private information exchange among multiple information systems can be realized, the delete right, the right to be forgotten and the extended authorization of the exchanged private information are still kept unsolved. Data under privacy protection schemes based on generalization, confusion, anonymity and other technologies cannot be restored due to fuzzy processing. These schemes are suitable for application scenarios such as single deprivacy and multiple deprivacy with a stepwise increase in privacy protection. However, since such schemes can harm data availability, privacy protection schemes with weak protection ability are often used in actual information systems. In addition, saving original data at the same time is also an option. Currently, there is a lack of description methods and computing models that can integrate privacy information and protection requirements as well as on-demand privacy protection computing architectures that can realize cross-system privacy information exchange, multiple service demands on privacy information sharing, and dynamic deprivacy in complex application scenarios.

It should be noted that there are still compliance pain points in the promotion and application of privacy computing. First, the authorization and consent mechanism still needs to be clarified when adopting privacy computing. Second, data security risks should also be taken seriously in the application of privacy computing.

Third, the realization of personal information subject claims in privacy computing applications still needs to be further explored. Next, we need to provide a complete set of privacy computing theory and key technology systems, including the privacy computing framework, formalized definition of privacy computing, principles that should be followed by privacy computing, algorithm design criteria, privacy protection effect evaluation, privacy computing language, etc. In regard to industrial practice and ecological opening, we also need to make more efforts to actually promote the practice and implementation of the series of technologies.

Although privacy computing realizes the dynamic balance between privacy protection and data cooperation at the technical level and plays an irreplaceable role in bridging data islands and releasing data value, what should be emphasized is that technology is the key means to achieve compliance, but a reasonable and scientific system is also an indispensable link in the data protection process. For privacy computing, both the regulation of legal systems and compliance with related laws, policies and standards can help realize data protection, which will be the precondition for data productization and commercialization.

Chapter 6
AI and Robot: Darwin and Rebellious Machine

As emerging technologies such as AI promote the development of technology to the paradigm of ubiquity and intelligence, it is particularly important to keep an eye on the development trend of technologies and human beings. As we have discussed before, modern technology not only brings benefits but also risks and uncertainties to human society, which makes people begin to worry about the alienation of human society and the loss of human essence caused by technologies. Therefore, how to promote the evolution of technology in the direction of "supporting" rather than "hijacking" human civilization is an issue that needs to be taken seriously. As Martin Rees, a professor at Cambridge University, said, "more technologies are needed to deal with global threats, but they need to be guided by sociology and ethics". Hence, this chapter mainly discusses some ethical issues related to the evolution and development of human society in AI, including AI consciousness, brain-computer interfaces, and automatic driving.

In the prevention and control of COVID-19, which began in 2020, technology ethics has become a topic that has aroused particular public concern, including ethical reflection arising from technology, ethical proposition for different groups in using technology, the exploration of the ethical connotation and ethical appeal on what makes a human being a human being. These propositions have philosophical implications and obvious practical significance. In studying them, we should be aware of the duality of AI ethics. One is the confusion and anxiety triggered by the risks it brings, and the other is that AI technology can also be regarded as an important way to solve problems. Thus, it can be seen that ethical issues are not only about the human future but also about the human imagination of their own ethical value system. Looking back on history, it is in the process of developing technology during which we constantly look back and reflect on the existing confusions that help us form a clearer judgment on future development, criticize and prevent risks, which is also the significance of our research on ethics.

6.1 Paradigm Revolution of Cognitive Science: Taking Consciousness Research as an Example

In the research process of AI, the research of cognitive science represented by neural networks has attracted much attention. The basic idea of the neuron model used in deep learning technology is generally consistent with the research on the linear threshold neuron model in cognitive science, and relevant research involves our understanding of basic intelligent problems such as cognitive science and mental models. In this section, we will discuss the research on mental models in cognitive science, especially the content related to neural networks.

In the 1970s, through the integration of interdisciplinary knowledge of psychology, anthropology, linguistics, AI and computer science, philosophy and neuroscience, cognitive science, an interdisciplinary discipline studying cognitive processes and minds, came into being. The core of this new discipline is to understand how information is represented, processed and exchanged in the nervous system, especially information about language, memory, reasoning, planning, decision-making, emotion, etc., and even study related to consciousness. Typical topics include the following:

(1) Is the nervous system an information processing system or a machine for meaning extraction?
(2) Since it is extremely hard to study consciousness, how can it be studied in a scientific way?
(3) How do neural networks store and employ memory?

As the topics above emerge one after another, the role of computers is becoming increasingly important, gradually forming an important branch, computational neuroscience. In the late 1970s, American scientist David David Marr proposed the computational theory of vision. He believed that information processing systems can be handled from three independent levels (theory, algorithm and hardware): computing objects can be determined through computing theory; algorithmic logic determines how to compute; and the hardware level settles what kind of computing structure can be used to support information processing.

It is the interdisciplinary research that enables us to better see the chimeric relationship between cognitive science and computational science and produce cognitive intelligence. Collecting, organizing and analyzing massive data is not only an important foundation of AI but also the premise of the next paradigm revolution of brain science. Of course, on the other hand, we should also see that it is unrealistic to only rely on existing computers to simulate the human brain (represented by the failure of HBP). Therefore, how to further study interdisciplinary computational neuroscience is a vital proposition.

Now, let us look at some basic conclusions of relevant disciplines. If we regard the brain as a neural network, two universal principles can be drawn. First, all our rich internal experiences and external activities are just the continuous action potential pattern of neurons. In other words, all experiences, such as external activities and internal mental models related to human behavior, can be explained by the potential control of neural networks. Second, most of our abilities to learn from past experience are largely due to the plasticity of synaptic connections among neurons; that is, through the study of neuronal connections, we can find the essential reasons for the brain to process information, and different local connection structures endow the brain with different functions.

Through the above two principles, an important idea is obtained: it is possible to realize artificial intelligence of cognitive function by simulating neural networks. Therefore, there are many models around neural networks, such as the famous McCullochPitts model and Hubel-Wiesel model, followed by the deep learning wave with great influence in this round of AI. In 2019, three pioneers in deep learning, namely, Yoshua Bengio, Geoffrey Hinton and Yann Lecun, won the Turing Award in 2018, and their main contribution is having laid the foundation for the development of deep learning through artificial neural network research.

What has to be mentioned here is the consciousness of AI. In fact, the great difference between weak AI and strong AI lies in whether it can produce some consciousness attributes through intelligence. Research on AI consciousness has deeper significance than research on AI consciousness. On the one hand, it simulates consciousness with machine, that is, the study of how to enable AI to produce consciousness. On the other hand, it is the research of manufacturing conscious AI systems. There are three theories in related fields, including "global neuronal workspace" (GNW), "integrated information theory" (IIT) and "life evolution theory".

"GNW" put forward by neuroscientist B.J. Baars believes that consciousness comes from the global workspace under the information sharing and exchange mechanism. It is the information dissemination within the whole cerebral cortex. How to understand the mechanism of how consciousness is produced in the physical system of the human brain is the key.

"Integrated information theory" originated from Gerald Edelman's "neuron group selection theory", which was formally put forward by Giulio Tononi. This theory holds that consciousness is related to the way and ability of the brain to integrate information. It is the fundamental attribute of any internal system with causal force acting on itself. The higher the degree of integration and differentiation of the system, the higher the degree of conscious experience it has. In other words, this theory denies that machines based on Turing machines can be conscious, but humans can construct conscious artificial intelligence based on the physical mechanism of consciousness.

"Life evolution theory" was proposed by Antonio Damasio. He believes that consciousness is an advanced way of regulating and managing life formed by life organisms in evolution. Understanding consciousness must start from understanding the essence of life and analyzing the self-balanced biological organization. Therefore, artificial consciousness is hard to create.

Finally, we provide a summative discussion on the consciousness of AI. N. Humphrey, a British psychosophist and psychologist, said, "Conscious feeling is the core of our existence. If we have not encountered this miracle, we are just poor lives in this insipid world." Through the above discussion, we know that the key to understanding consciousness is to understand the nervous system in the brain.

Consciousness is a unified neural model that brings experience content and self together from the most basic to the most complex level. A large number of scientific studies have shown that many human mental functions are completed unconsciously. Therefore, no matter what technical means (EEG, PET, fMRI or optogenetics) are adopted, the core problem of neurobiology of consciousness is to study the neural mechanism regulating and supporting the state of consciousness by studying the difference between conscious and unconscious brain states.

Moreover, the metaphysical study of consciousness is also a crucial field, such as the study of dual-aspect monism. Relevant theories believe that the world is composed of an ontology, and each instance of ontology contains two different and mutually reduced aspects, the physical level and the experience level, which solves the contradiction between the conceptual assumption and the phenomena of consciousness of scientific materialism and dualism.

The content above is our discussion of cognitive science research, which seems to have little to do with the current development of AI. However, it is actually an important development trend of AI. Since the 1990s, the science of consciousness has developed continuously to address issues such as the origin and evolution of consciousness, free will, and the problems of other minds. We have seen the diverse vision of this field, and the future research of AI is universal. Moreover, as the research of consciousness is also closely related to social science, ethics and public policy, how to study AI ethics and responsible AI is also importantly linked with cognitive science.

6.2 Human-Computer Symbiosis Under a "Brain-Computer Interface"

After discussing brain and cognitive science, let us discuss the impact of the technology trend of the "brain-computer interface" (BCI). Of all the black technologies, none will completely overturn our cognition of human beings such as BCI.

Its subversive character manifests itself in trying to replace the most important cooperation tool in the process of human evolution—language. A brain-computer interface is creating a new communication interface and mode by directly communicating with the outside world through the brain, which will not only bring changes in the way we communicate but also lead to a series of derived abilities, such as conscious manipulation of machines, comprehensive improvement of brain examples, replacement of the human body with mechanical bones and so on. This section discusses the ethical issues of "man-machine symbiosis" in combination with the development of BCI technology.

Let us first look at the specific application of BCI from four aspects. The first aspect is the most basic one, "application repair", which refers to how BCI technology can repair physical functions. For example, this application can enable the paralyzed to stand up and walk again or enable the blind to restore vision. Since this is the original intention of BCI, it is also the most deeply studied application in laboratories. The repair here includes not only the replacement of physical functions with machines but also the restoration of their own physical functions.

The second application is "improvement". Scientists can use brain-computer technology to improve our mental state by collecting EEG signals and analyzing brain states, such as improving attention and sleep quality and even stimulating flow experience. This is the most likely commercialized aspect of brain-computer technology.

The third aspect is "enhancement". Elon Musk once argued that BCI is the equivalent of adding a new structure called "the third digital layer" to the human brain. The new structure will dramatically increase our intelligence and surpass our biological limits, realizing the integration of human beings and machine intelligence.

The fourth aspect is "communication". BCI can realize the direct interaction between brains, forming a "lossless" brain information transmission mode, namely, the direct interaction of neuron groups. This application can provide conditions for the future super brain through "emergence".

Of course, many of the above studies are still in a very initial stage. What we are mainly discussing here is the brain science and cognitive science revealed by brain computer interface technology. Through the study of animal brains, scientists have found cognitive maps that can help them establish different internal models, track spatial physical locations and form multiple psychological processes (memory, imagination, reasoning and abstract argument). The research of Edward Tolman, an American scientist, shows that the brain works like a telephone switchboard that is only responsible for maintaining those it considers reliable signal connections. These signals are either dialed from sensory organs or sent to muscles by the brain. It is through continuous "learning" that the brain can form different "cognitive maps" to realize related complicated functions. His research is actually similar to the BCI theory in effect we discussed here.

From the perspective of historical development, BCI technology formed in the 1970s is an interdisciplinary technology involving brain science, cognitive neuroscience, computational neuroscience, etc. In 1973, Vidal first proposed the concept of BCI. He believes that the purpose of its application is to help patients with dyskinesia control external devices and create a new approach for them to interact with the external environment. In other words, BCI technology was originally invented to realize the connection between patients with neuromuscular disorders and the outside world.

It was not until the International Conference on Brain-Computer Interface 19 years ago that scholars formally defined BCI technology: brain-computer interface technology is a communication system that does not rely on the normal output pathway composed of peripheral nerves and muscles, and it can directly provide a new information exchange and control pathway for the brain, creatively helping the brain interact with the external environment or devices directly.

For the application principle of BCI technology, neuroscience holds that the activities of the individual nervous system will change after being stimulated by the outside world, and these changes will trigger the next action. Hence, certain means are used to detect, classify and identify these electrical activities of nerves as signals of specific actions and then program them in computer language to convert the thinking activities of the individual brain into computer instructions driving external devices to realize the direct interaction between the individual brain and external devices without relying on the surrounding nerves and muscle tissues. BCI technology has realized direct communication between brain thinking activities and the outside world and has even been able to control the surrounding environment with this principle so that human beings have the super ability to control "action" with "thought". Similar to communication and control systems, the general BCI system also consists of three parts: signal acquisition, signal processing and an interactive control module. With this system, human beings can directly achieve interaction and feedback between the brain and the outside world.

BCI technology was first applied to medical detection and rehabilitation medicine. With the deepening of the understanding of nervous system functions and the development of computer technology, BCI technology is gradually growing mature. Its application field is expanding, and its research also shows an obvious increase. Let us discuss a few typical application scenarios of BCI technology.

The first is the field of education. As a significant result of brain science research in recent years, BCI technology can read and convert brain bioelectric signals into "action" through signal conversion and computer programming, achieving more natural man-machine interaction, which brings new opportunities and challenges to the application of educational technologies.

The second is the medical field. As BCI technology can directly help the brain interact with external devices and skip the conventional brain information output path, it has broad application prospects in the medical field. The output of the BCI system can replace the natural output lost due to injury or disease, which is helpful for patients with nervous and muscular paralyses to express their willingness and conduct rehabilitation training.

Rehabilitation training is a typical field of application. For example, the research team of Xi'an Jiaotong University designed a set of "brain controlled human-computer interaction and rehabilitation training systems", which can help trainers better monitor the intensity and effect of the training and adjust the training plan according to the feedback through a new brain controlled rehabilitation mode realized by BCI technology.

In the field of intelligent prosthetics, if people lose their limbs, they tend to wear artificial limbs with fixed structures that normally cannot realize the functions of the original ones. The artificial limb connected with the patient's brain through BCI technology can achieve autonomous control after a period of training. A Harvard research team has created a semipublic welfare project BrainRobotics to manufacture intelligent artificial limbs for the disabled. The products developed in the project can help the disabled control prosthetics with their own thoughts.

In addition, there are practical applications in other fields. For example, Rythm, a U.S. company, has designed a head-mounted product called "Dreem" that can identify the wearer's sleep mode with the help of BCI technology. To improve the sleep quality of the wearer and help him/her maintain deep sleep for a long time, some auditory or sound stimulation is needed according to different sleep modes.

Due to the importance of BCI technology, the United States and other countries are actively developing corresponding technical research results. For example, in March 2016, the Defense Advanced Research Projects Agency (DARPA) launched the next generation noninvasive neurotechnology program (N3) and announced the selection of 8 research teams in April 2017. The program aims to develop nerve stimulation methods to activate "synaptic plasticity" and establish training schemes strengthening or weakening the connection between two neurons to accelerate the acquisition of cognitive skills and improve the effect of skill training.

As before, let us discuss the technical ethical risks or social risks faced by BCI. In addition to the common problems of emerging technologies, including security, fairness, privacy and discrimination, BCI technology has also raised some special problems. Since the technology involves the brain, the most important organ of a human being, it can trigger the problem of "brain control", which can negatively affect human self-identity, autonomy and personality.

In regard to privacy, the long-term recording and decoding of a single brain signal can dynamically monitor the brain state and "intention" in real time to reach optimal analysis. However, these data are all about personal core privacy and spiritual information. Protecting the privacy and integrity of brain data is the most valuable and inviolable human right. Therefore, we need to attach great importance to the use of these data to protect users' privacy when developing relevant technologies. For example, whether a person's intelligence data can be used in various recruitment, employment and promotion scenarios. Whether a person's personality, political orientation and sexual preference can be used as the basis for selection and employment. Whether a person's intelligence level, the health condition of the brain and the probability of developing some brain diseases can be used to determine the price of services (such as insurance, training, etc.).

The second worry is the proposition of social equity caused by BCI technology, such as the inequality of material wealth, different living conditions and medical levels as well as the resulting difference in life expectancy of different people. All these differences with great impact are striking and intuitive and are of great concern to society. With the development of BCI technology and cognitive enhancement techniques, such technology can only be used by the majority owing to reasons such as price, technology regulation, market regulation, etc., resulting in more obvious unfairness.

For example, if those few who have access to intelligent enhancement devices perform better in exams and work and gain advantages in income and social status, inequality between people will increase. Studies have found that intelligence is an important factor that affects a person's performance in many fields and even social strata. Thus, the cognitive inequality caused by the exclusive use of cognitive enhancement technology by the minority will cause a deeper social gap and may further aggravate social inequality, which is hard to reduce through existing economic and administrative approaches such as taxation.

The last concern is about identity. As the object of BIC is the human brain, the technology can directly change the core of human beings, including cognitive ability, personality characteristics and even self-concept, which may bring very profound self-identity alienation. The identity and self-concept of human beings with continuity and consistency need a long time to establish. They are stable and change slowly. Once these slow patterns are broken by BCI therapy, so many and so fast changes will interrupt self-identity and have a negative impact on human beings.

In rehabilitation medicine, the motivation of human subjects to participate in treatment is to have complete body parts, realize effective communication and movement, and eliminate their patient state. However, if the treatment produces remarkable results in a short time, identity difficulty may occur due to such ideal and sudden change. The same is true of medical ethics, such as disability treatment and cosmetic surgery, which will also change self-concept. The change may be more evident in BCI. Imagine that a person who was depressed two days ago becomes extremely active and outgoing after treatment, which will not only surprised others but also confused himself/herself.

More importantly, BCI also damages self-identity owing to attribution, interfering with people's sense of self-identity and shaking the core hypotheses of self-identity and personal responsibility. In a research of scientists, a patient who had suffered from depression for 7 years reported in a focus group that he was confused about "who he is" and began to doubt whether the way he interacted with others was controlled by himself or the device he wore after receiving brain stimulation treatment. This deep-seated self-identity confusion can cause much confusion and deeper emotional disturbance.

Considering the series of social impacts brought by BCI, researchers must think about how to run the sense of responsibility through the whole innovation process of BCI technology to avoid risks to the greatest extent and better benefit humankind.

Kevin Kelly, the author of *Out of Control*, believes that human beings have experienced four cognitive arousals. The first time occurred when Copernicus put forward the heliocentric theory. We pulled the earth down from the altar. It turns out that the earth was not the center of the universe. We reshaped the relationship between human beings and the universe. The second time is when Darwin put forward the theory of evolution. We pulled God down from the altar. This proves that people have evolved from monkeys. What about the third time? The philosopher Freud broke the cognition between human and self with the theory of self-consciousness. The last is that human beings rerecognize their relationship with machines, and more cyborgs may become legal citizens in the future. In my vision, machines will be given lives, and the combination between human and machine intelligence will become increasingly close. This is also the best time for us to readjust the technical ethics framework and achieve the synchronous evolution of human social ethics and technologies.

6.3 Automatic Driving: "Survival of the Fittest"

After discussing BCI technology, let us discuss a widely recognized technology, "automatic driving". From the end of 2020 to the beginning of 2021, intelligent networked vehicles are surging, automobile intellectualization and networking continue to heat up, and new products, new players and new technologies have emerged. 2021 is called the first year of the outbreak of automatic driving.

The definition of automatic driving technology is divided into two systems, L0-L4 and L0-L5, according to the standards issued by NHTSA (National Highway Traffic Safety Administration) and SAE (Automobile Engineering Association). It is a relatively long process for automatic driving to upgrade from L2 to L5 in the latter system. The automatic driving path in most automobile factories is upgraded from L2 to L5 step by step. The further the development of automatic driving technology, the higher the requirements for sensing components such as radar and camera, algorithms, data support, etc. For auxiliary driving at L2 and above, strong data support, such as ultrawide data transmission and processing and the reaction and processing ability of radar and camera if an event is sensed, is needed.

Automatic driving integrating technologies such as artificial intelligence, communication, semiconductors and automobiles involve a long industrial chain and have a huge space for value creation. It has become a crucial market for cross-border competition and cooperation between the automobile industry and science and technology industry of countries worldwide. At present, automatic driving has come to a historical junction. From Baidu Robotaxi to Yutong WITGO and from Zhengdong New District to Chongqing Yongchuan, China's commercial vehicles and open roads equipped with L3 and L4 high-level automatic driving technology are gradually scaling up. However, with the maturity of automatic driving technology and increasingly extensive market application, the resulting ethical controversies are also the most heated debates of AI.

Tesla have recently been confronted with frequent accidents involving self-driving cars. The alarm of a number of L2 automatic driving accidents is still sounded in our ears. In addition, greater challenges, such as bad weather, temporary control, traffic intersections, falling objects on the road or reflection, even make safe driving more challenging. When dealing with all-weather and all-scene driving conditions, automatic driving still has a long way to go.

Let us first review the governance events in automatic driving. In June 2017, the German Federal Ministry of Transport's Automated Driving Ethics Committee submitted the *Report on the Ethics of Autonomous Vehicles*. The document pointed out three macro ethical requirements: first, the primary goal of automatic driving is to promote the safety of all traffic participants and to lower the risks of all traffic participants on an equal basis; second, it believes that accidents occurring under an automatic driving system do not exceed ethical boundaries; third, although automatic driving technology can reduce risks and improve road safety, it is not proper in an ethical sense to mandate the use and popularization of this technology by legal order.

The report shows that Germany emphasizes that technology should follow the principle of individual autonomy and should not restrict individual freedom of movement. The German ethics report acknowledges that highly automated systems with anti-collision functions can reduce risks and enhance safety, which is encouraged at the social and ethical levels. However, if this technology is forced to use by law, some moral disputes may appear. For example, people will be subjected to technology, and the status of humans will be derogated and debated. Of course, the design choice of driverless cars also faces pressure and resistance from the driver group. With the maturity and popularization of driverless technology, disputes over whether human driving will be replaced by automatic driving and whether automatic driving takes precedence over human driving have also raised concerns.

Before discussing the ethical disputes of automatic driving, we should first sort out the difference between automatic driving and unmanned driving. According to the classification standards of automatic driving vehicles proposed by NHTSA and SAE, L0 refers to "manned driving", also known as "no driving automation", L1 refers to "drive assistance", L2 refers to "partial driving automation", L3 refers to "conditional driving automation", and L4 refers to "driving automation". Among them, L4 "driving automation" is divided into "high driving automation" and "full driving automation" according to the standard of SAE.

For the sequences before L3, the human driver needs to be ready to serve as a "driving scene supervisor" and "emergency overtaker" at any time, which may not be much easier than manned driving and sometimes with less driving fun. Therefore, as you see, "unmanned driving" is only applicable to L3 and above level vehicles in the broadest sense. The "AI-assisted driving vehicle" before L3 can only be regarded as "automatic driving" rather than "unmanned driving", which means the former is a concept including the latter.

In terms of the technical path, the development of autonomous driving technology was initially developed on the basis of the combination of traditional automobile technology and artificial intelligence. The first solution is constant speed cruise control, which enables vehicles to cruise automatically at a certain speed and under certain circumstances. What came next was the development of automatic anti-lock breaking technology (ABS), which can limit braking when emergency braking does not work while driving. The second is lane departure warning technology. If the vehicle deviates from the predetermined route, it will give a warning. Then, come automatic parking technology that allows cars to return to and from garage and wait to be used when receiving instructions. Nevertheless, these are not fully autonomous vehicles. A real automatic driving vehicle means that the user does not need to control the vehicle at all. The automatic driving system automatically follows owners' instructions and completes delivery instructions.

The British carmakers predict that by 2027, most of the cars produced in Britain will be equipped with at least L3 automatic driving technology. Most of the automatic driving cars currently being developed, manufactured and tested are L3. By 2030, over 25% of British vehicles can achieve full automation. All of them have no steering wheel at all. When someone gets on an autonomous car and gives instructions to it, it will automatically drive him/her to the designated destination. In other words, in just over a decade, a quarter of British cars will be fully autonomous. Hence, autonomous vehicles have very promising prospects. The major auto manufacturers are now fully committed to the technology of automatic driving. China's auto driving vehicle has also been doing well. The Automation Research Institute of Beijing Institute of Technology has developed dozens of self-driving cars that can drive autonomously in formation. It is expected that the market potential of semiautonomous and fully autonomous vehicles will be considerable in the next few decades. In 2035, China will have approximately 8.6 million autonomous vehicles, of which approximately 3.4 million will be fully autonomous.

However, with the continuous upgrading and popularization of autonomous vehicles, ethical and legislative difficulties in road traffic are standing in front of researchers. Although many companies, such as Google, BMW, Audi, Volvo, Mercedes, and Nissan, are currently developing automatic driving technology, all companies involved in automobiles have noted that automatic driving cars are safer. Most experts also agree that introducing automatic driving vehicles into society will reduce traffic accidents as well as traffic fatalities. However, due to software technologies, algorithms, hardware equipment and other reasons, advanced vehicles can never be one hundred percent safe. Nevertheless, we need to identify who should be responsible for the accidents caused by it.

Relevant responsibility can be divided into causal responsibility and moral responsibility. Causal responsibility mainly explains the consequence of each action. For example, lightning may set off forest fires; we can say that lightning is a causal responsibility instead of a moral responsibility. Some ethicists deem that there is a "responsibility gap" in self-driving traffic accidents; that is, it is not clear who should bear the moral responsibility when autonomous vehicles cause injuries or losses.

Some scholars have studied the responsibility of automatic driving car accidents and put forward a "responsibility field" for self-driving traffic accidents. Carmakers should not be responsible for accidents caused by automatic driving unless they know their products are defective and never fix them before selling due to the high cost of fault correcting. In addition, the unrepaired defect caused by the accident is another indispensable condition. Users cannot fully predict the actions of their automatic driving cars, so they lack the necessary reaction time to avoid accidents. Thus, they cannot be held responsible for such accidents, nor do they need to be liable for the corresponding damages.

Legally, the key to such a liability gap lies in the fact that autonomous driving vehicles do not have the conditions to be the subject of legal qualifications if they are involved in traffic accidents. From the viewpoint of legislative logic, if we want to clarify responsibility attribution when automatic driving causes damage, we should first determine whether AI can be given the status of legal subject. In regard to this proposition, there are three schools of thought in the jurisprudential circle, namely, "subject theory", "object theory" and the view of regarding automatic driving vehicles as a special existence between "subject and object". We will talk about them in detail in the following pages. The "subject theory" takes an inclusive and open attitude toward autonomous vehicles. It believes that autonomous driving vehicles that can make autonomous decisions have the similar "rational", "free will" and thinking process of human beings, which conform to the essential condition of the "free will" of legal subjects and should be given a subject position. The "object theory" denies the status of the legal subject of autonomous driving cars. It supposes that automatic driving cannot be equated with manned driving and that there is a huge gap between them.

However, quite a few scholars believe that even with the status of legal subjects, the automatic driving vehicle does not possess the legal personality of the human being; it cannot satisfy the requirements of substantiality and value of fictional subjects. Thus, it cannot bear the actual responsibility, which makes the human being the ultimate responsibility bearer, although automatic driving cars have the status of legal subject in a sense. At this point, it makes no difference whether automatic driving cars are given the status of legal subject or legal object in terms of legal effect.

From the perspective of ethics, automatic driving may also cause many ethical plights. In the study of ethical issues of autonomous driving, the "plight" faced by machines in moral decision-making is a classic topic of ethical debate. People are always confused about how to make choices when harm is inevitable.

In the "trolley experiment" and other traditional ethical experiments, automatic driving often faces the dilemma of decision-making. Either choice will inevitably lead to ethical questions on machine decision-making. The focus of this contradiction is the different standards of moral choice caused by utilitarianism and virtue ethics. However, most moralists hold the principle of universal law and tend to "not rule out the happiness of others". The conflicts between the two standards lead to difficulty in formulating the code of ethics for autonomous driving. Utilitarianism ignores personal disputes, and the damage may not be calculated. When a driverless car hits two electric motorcycle drivers. One of them is wearing a helmet, and the other is not. The one without a helmet is likely to be hit, which is obviously unfair and may lead to deeper moral hazards.

In addition, the moral decision-making of machines also involves a realistic problem—who will supervise the system of automatic driving vehicle to ensure that it carries the correct moral modules? According to the principles of the Department for Transport in Britain, if the alliance of manufacturers of automatic driving is allowed to carry out supervision, the passengers directly related to this technology have no voice. This is definitely wrong. Since ordinary passengers have no relevant technical and moral knowledge, it is difficult for them to conduct effective supervision. At this time, is it necessary for the government to undertake the supervision of citizens? As every hospital should set up its own ethics committee according to law, should automobile manufacturers follow suit and set up their own ethics committee? If these problems are not handled well, some ethical risks may occur. For example, the implementation of the moral module may not be well supervised.

Moreover, decision-making autonomy has also attracted public attention. Once people accept the auxiliary driving function of automatic driving, they will inevitably rely on the intelligent driving system. Such dependence will continue to increase with the upgrading of the intelligence level of the system, forming ratchet effects. That is, "consumption is easy to rise but difficult to fall". Automatic ethical decision-making may lead to the risk of completely breaking away from the ethical decision-making circle, which may pose a direct threat to users' moral autonomy. On the surface, this is a problem of right shift between man and machine, not to mention the lack of a broad consensus on the degree of people's willingness for the right shift. From a deeper level, rights are shifted among automatic driving users and developers, mobile Internet service providers and navigation and positioning service providers. Driven by commercial interests, the right shift of autonomous driving users may not be voluntary or transparent, so it is worth further discussing setting certain principles to protect consumers' rights.

As the intelligent decision-making of automatic driving depends on big data, it inevitably brings privacy problems. We should disclose more relevant data of companies to drive safely, but it seems that this disclosure should be restricted for the purpose of protecting copyright and trade secrets. Similarly, more data should be shared for users' traffic safety, but less data should be collected for users' privacy. Moreover, the data from the self-driving vehicles' detection of the road are owned by the company. However, if some national security facilities are detected by autonomous vehicles, national security will be undoubtedly affected. Privacy is not an independent issue; it involves other sensitive issues. Additionally, we need to consider the following questions. How much data should be disclosed and to whom? How can data privacy protection principles be set?

To solve the above problems, we can start from the following aspects.

First, we should promote the training, education, publicity and popular science of automatic driving to prevent the public from supporting or resisting the technology blindly. Social public institutions, especially publicity departments, should assume the responsibility of popularizing the principles and development significance of automatic driving technology to the public and dispelling the public's fear or adulation of automatic driving through various channels and innovations, transferring the role of the public from the technology beneficiary to the mutual benefit beneficiary. Moreover, universities and other scientific research institutions also need to be fully aware of their mission and responsibility. They also need to strengthen the learning and thinking of examining autonomous driving from a social and humanistic perspective in the process of education while delivering automatic driving-related talent through scientific training on the basis of industrial characteristics.

Second, we should respect and safeguard the rights and interests of special groups and build automatic driving ecology with warmth and love. Social governance participants should fully realize that the digital public transportation represented by automatic driving vehicles is essentially based on unfair or even differentiated social group rights and interests. This requires public management institutions to innovate social governance initiatives and take full consideration of and guarantee the rights of vulnerable groups to enjoy public resources through various methods. In addition, we should innovate social governance thinking and replace the result-oriented intelligent transportation construction logic with people-oriented humanistic care, adding more warmth and care to the travel experience in the future.

Third, we should avoid technology abuse and personality alienation and enrich the application scenarios of automatic driving. Technology abuse is mainly manifested in two aspects. One is the excessive application of autonomous driving, such as the unreasonable application of automatic driving in some highly complex scenarios. The other is the abuse of personal information caused by the use of automatic driving cars. To avoid the excessive use of automatic driving itself, public management institutions should clarify the application scenarios of automatic driving and scientifically compute its risks and benefits. According to the principle of classification and grading, we should systematically plan the application methods in different scenarios and effectively evaluate the cost, actual benefits and potential risks of automatic driving in different scenarios. Particularly, when automated driving is applied to public management, it is necessary to consider the individual's cognitive cost of human–machine specification and machine operation mode and continue to bring well-known social norms serving part of the basic social standard space into full play to avoid the contradiction and conflicts between the general public and automatic driving vehicles.

Apart from the above points, we also need to provide more in-depth guarantees at the technical and industrial levels, such as paying attention to technical escort and supply chain security. The key technologies of automatic driving are mainly based on the combination of intelligent technologies, including mobile communication, computer vision, reinforcement learning and vehicle engineering control.

In terms of the development of AI in China, three major technological breakthroughs need to be made:

The first is the implementation of hybrid-augmented intelligence, including systematic research on corresponding computational reasoning, model and knowledge evolution, especially the application and breakthrough of related technologies in visual scene understanding. The next is research on active vision systems based on superhuman vision and the basic theory of visual perception-oriented autonomous learning. The last one is the technology application and breakthrough on a complex perception reasoning engine, city level system platform and the related new architecture chip capability. In addition, the following specific measures can be taken in the AI industry chain.

1. Enhance AI core competence: the R&D and application layout of AI general technologies;
2. Promote platform construction, build city-level intelligent transportation infrastructure platform, and promote the landing of autonomous driving technology in complex smart city scenes;
3. Strengthen the ethical code design of the logic system of top-level building;
4. Improve the interpretability of algorithms related to automatic driving and the ethical constraints of independent decision-making;
5. Ensure that crisis situations can be taken over and blocked.

Regardless, it is certain that the birth of driverless cars will have a profound impact on related industries, including urban construction, automobile manufacturing, energy and the human world, just as cars appeared in people's vision like beasts over 100 years ago. We do not need to be overly afraid when encountering various unsolved ethical problems relevant to automatic driving. Because ethical problems are not exclusive to automatic driving, they cannot come to an end through automatic driving technology or become an obstacle to the application and promotion of this technology.

We might also maintain an open attitude toward automatic driving, similar to Darwin's theory of "survival of the fittest animals in natural selection", let technology and the market decide who will dominate the road ahead.

Chapter 7
Virtual World Under AI: Augmented Reality and Deep Synthesis

The relationship between people and digital information has been changing all the time. On the one hand, people can obtain an enormous amount of information from texts, images and videos in various ways. On the other hand, from keyboard, mouse, touch screen to voice, the interactive modes of information are getting closer to the five natural senses of the human being. The popularity of the virtual world will promote the evolution of the relationship once again. The presentation of information will move toward the integration of virtual and real information, and the interaction of information will step toward a totally new form driven by technologies such as image recognition and gesture recognition.

From a macro perspective, as the global economic situation has remained complicated and changeable, COVID-19 has changed the way people live and produce. Information consumption and industrial digital transformation are playing an increasingly prominent part in human production and life. The typical and innovative application scenarios integrating new technology paradigms with the new generation of information technology continue to expand and play a key role in human production and life.

Taking AI as an example, AI mainly aims to help human beings achieve smarter decision-making and rapid response in unsupervised environments. The support does not stop at data processing and content distribution in a single scene but evolves and upgrades from presenting data itself to showing the structural relationship among data, which strengthens the fitting degree between real life and the digital world, thus pushing humankind to build a virtual world under AI. Virtual reality is one of the most critical forms, especially with broad application scenarios in mass consumption and vertical industries, and VR and AR are in their golden period of industrial development.

In the meantime, with the rise and evolution of AR/VR, cloud computing, AI, 5G and other technologies as well as an increasingly elaborately constructed virtual world, the concept of "**Met aver**se", the ultimate form of the Internet, is booming.

This chapter will start with **augmented reality** and **deep synthesis** because they are making a virtual world highly interconnected with the real world. A typical application, digital human, was chosen as an example. The chapter will lead you to explore the ultimate future form of the virtual world—"Met averse", which integrates diversified digital technologies to study the risks and opportunities we face. AI has been used not only to create human intelligence but also to create cyberspace. The ethical issues we are discussing also need to be expanded to a broader sense.

7.1 AI+AR: A New World Under Augmented Reality

Augmented reality, also known as AR technology, is an information technology derived from the limits of the complete integration of virtual reality and realistic environments. In 1990, the Research Institute of Boeing took the lead in proposing augmented reality system theory. After decades of development, AR technology has been continuously updated and iterated. Currently, many better mainstream augmented reality frameworks have been derived, such as ARtoolk and Vuforia. **The technology is designed to generate virtual objects that do not exist in the physical world with the help of computer vision technology and AI technology and to accurately "place" virtual objects in the real world**. AR put the virtual world on the screen of the real world and interacts with it. Through more natural interaction, AR presents a richer perceived new environment for users. The key distinctive feature between augmented reality and virtual reality is that the former is realized by precisely superposing virtual information onto the real world to achieve the combination of virtuality and reality.

The principle of AR technology is to capture the images of the real world through a camera. People can input commands into equipment through voice and gestures. Then, computers use computer vision technology to understand the surrounding environment and conduct recognition and interaction at the same time. The outcome will be processed by the rendering engine, and finally, the technological result will be output by display, achieving the effect of the fusion of reality sense and virtual experience.

The core of augmented reality lies in human-computer interaction, and as the underlying technology of AI, AR plays a crucial role. In other words, AR is actually a visual representation and interaction method of AI. The mainstream augmented reality products in the market are divided into three categories: head-mounted displays, handheld mobile terminals and spatial displays represented by PCs and HUD flat panel displays (FPDs).

In terms of market penetration, PC displays and mobile terminals are slightly higher than the head-mounted displays represented by AR glasses. However, since AR glasses have broken through the limitation of the screen, the whole physical interface may become the interactive interface of AR in the future. Near-to-eye screens such as AR glasses may be the future development direction of augmented reality hardware.

Game products developed based on AR and computer vision technology currently make AR accessible to ordinary families. Nintendo, the world's leading entertainment manufacturer, cooperated with Niantic Tokyo Studio to develop *Pokémon GO*, an AR handheld game based on the IP of *Pokémon*. Since its launch in 2016, it has created multiple download and activity records. It topped the IOS free and best-selling list in Australia, New Zealand and the United States, and Nintendo's share price soared by 9% after *Pokémon GO* was listed in these three countries for only one day in July 2016. More than a month later, the game had set five Guinness World Records, including the highest revenue in the first month (US $206.5 million), the most downloaded games in the first month (US $130 million), the most installed games in the first month (70 countries/regions), the most revenue in the first month (55 countries/regions) and the fastest revenue growth that exceeds handheld game revenue (over $100 million in 20 days) by $1 billion.

Pokémon GO has been maintaining a high level in business value since its release. Especially under the influence of COVID-19, the classic is popular with the game player under the family segregation policy, resulting in a total growth of 11% in game player spending in the first ten months of 2020 compared with that of 2019, up 30% over the same period.

Specifically, the driving force of AR industrial upgrading comes from the continuous breakthrough of various technologies. The technology system of AR can be divided into the following four categories: **perception, modeling, presentation and interaction**. Perception mainly captures the position and movement of eyes, head, limbs and other body parts. Modeling is used to establish digital models from multiple aspects, including geometry, physics, physiology, and behavioral intelligence. The presentation aims to realize more vivid and lifelike digital virtual images integrated with reality based on technologies such as 3D displays, spatial audio, and near-to-eye displays. Interaction is designed to convey correct instructions to the computer through the sense of touch, voice, body feeling of humans. Apart from technologies, AR requires a stable and efficient operating system that should support multi-tasking, synchronous positioning and modeling, 6DOF and efficient rendering. For the AR technologies above, the key lies in the technologies of positioning and interaction, which are mainly realized by external laser positioning, internal/external image processing positioning, spatial scanning, whole-body motion capture, eye tracking technology, natural interaction technology, etc.

Let us first look at the applications of AR in some specific fields. As AR technology with two-wheel driving forces, policy bonuses plus market education, is being implemented in education, the teaching mode is upgrading from passive acceptance to independent experience. For curriculum content that is difficult to remember, practice and understand in traditional teaching processes, AR can help improve teaching quality and the effect of industry training.

In the field of education and teaching for the masses, Edgar Dale's "cone of learning" theory assumes that the more learning situations people participate in, the stronger their memory will be, and the knowledge and experience gained through first-hand experience will be much more effective than through traditional teaching and training methods (text symbols, audio broadcasting, static pictures, etc.). AR can help students gain "hands-on" experiences that are difficult to realize in the real world by letting them interact with virtual objects, complex phenomena and abstract concepts to stimulate learning, improve attention levels and knowledge retention rates and reduce potential security risks. Moreover, AR can improve teaching efficiency and unleash the creative potential of the next generation of information technology.

Enterprise-oriented skill training, according to the targets of enterprise training of all sorts, can be divided into task-based training, multiperson collaborative equipment and facility training and soft skill training based on AI. Unlike standard courses such as K-12 in the education market, long-tailed AR training across vertical fields in the enterprise market is highly customized. Innovative AR application requires a clearer return on investment. For instance, in the field of factory vocational education, the AR platform provides workers with multiscene vertical industry virtual reality training solutions such as equipment operation drills, process simulations, safety accident recovery and explanations of structural principles, intelligent inspections and skill assessments to improve the training effect and efficiency.

AR can be regarded as the new carrier of the new information consumption mode in the field of cultural entertainment. Traditional cultural entertainment experience features limited interaction, insufficient socialization and a single experience form. Virtual reality supports integrated, shared and immersive digital content and services, focuses on the integration and innovative application of information technology, creates an upgraded version of information consumption, and cultivates new growth points in medium- and high-end consumption areas. AR is mainly used in entertainment and recreational fields, including supermarkets, tourism, social networks, games, drama and live broadcasting of events.

For example, AR navigation can achieve "centimeter" precision in intelligent business complexes. It can realize high-precision spatial positioning of indoor and outdoor areas by reconstructing a large-scale 3D map through visual perception technology by scanning the surrounding environment from the underground parking lot to any floor in a mall with mobile phones and employing AI technologies such as simultaneous localization and mapping (SLAM). As all road signs and instructions, for example, where to turn and where to take an escalator, are directly concluded in real scenes, "AR arrow", the landmark indication, will considerably indicate the directions to the destination whether you are shopping for trendy items or participating in the latest theme activity. AR can make your travel convenient and pleasant because you don't need to identify directions anywhere anytime.

Let us look at more specific business cases next. In 2019, the AI+ AR Museum Project "Chimebells from the Tomb of Marquis Yi of the Zeng State", jointly launched by Sense Time, Hubei Provincial Museum and Suzhou Yunguanbo, made its debut in the "AI+ Culture and Arts" exhibition area of Sense Time in 2019. Visitors could play the virtual chimebells created by AR technology with their hands instead of a hammer by pointing the cameras of their mobile phones, tablet computers and other devices at the real object or the picture of this set of musical instruments. By this means, they could experience the fun of playing ancient music and feel the charm of national treasures in person.

The project has employed three AI+ AR algorithms: first, high-precision positioning was adopted to realize accurate virtual-real fusion between the virtual scene and the real scene; second, SLAM technology was used to ensure that the virtual

7.1 AI+AR: A New World Under Augmented Reality

chime can be stably displayed in the real scene so that the audience can feel that the real chime is just there no matter which angle they choose to appreciate it; third, through the dense geometric reconstruction of three-dimensional space, the chime had a real occlusion location relationship with the surrounding environment. Moreover, apart from visual "reality", the bell sounded by the audience also came from the original sound recorded when the chime was tolled.

Looking ahead, with the empowerment of AI technology, AR will realize more accurate perception and recognition of scene environments, objects and people and further achieve the perfect integration and seamless connection between virtual information and the real world. The authenticity, content and application scope of experience will be explosively improved. Meanwhile, by more flexible and free connection between devices, more creativity will come out in multipeople collaboration. The unified AR cloud platform will realize multipeople space sharing and create a new information world. In the future, the development and evolution of virtuality and reality will no longer be a simple superposition but an organic integration. Under the innovation system of cloud computing, network, edge computing, terminal, application and human integration, we should reconstruct the existing system architecture and trigger a leap in the industry to produce a number of new technologies, new products, new standards, new markets and new business models that will be redefined and iteratively optimized under the framework of deep integration and innovation.

As such, the concept of "Metaverse" came into being. "Metaverse" can also be called the new generation Internet, where reality and virtuality interact. VR/AR will become the technical carrier of the new generation Internet.

The concept of "Metaverse" originated from *Snow Crash*, a science fiction written by Neil Stephenson. The story describes a world in which people interact with software as virtual avatars in three-dimensional space. "Metaverse" is considered the "ultimate form" of the Internet.

Conceptually, the word "Metaverse" is made of "meta" and "verse", in which "meta" stands for transcendence and "verse" represents the universe. They usually represent the concept of "beyond the universe", an artificial space operating in parallel with the real world, when put together. Looking back on the history of the Internet, from PC LAN to mobile Internet, the immersion of Internet use is gradually increasing, while the distance between virtual and reality is gradually shortened. Following the trend, the "ultimate form" of the Internet, "Metaverse", is formed when immersion and participation reach their peaks.

Technically, based on the traditional Internet, metaverse space will be supported by many independent tools, platforms, infrastructure and protocols because it puts forward higher requirements for immersion, participation and sustainability. As AI, AR, VR, 5G, cloud computing and other technologies are growing mature, metaverse space is expected to gradually move from concept to reality.

According to our bold imagination, metaverse will have the following four core attributes: the first is "synchronization and skeuomorphism", which means the virtual space and the real society maintain a high degree of synchronization and interoperability, the interactive gap between human and computer will become increasingly smaller, and the virtual world will be close to reality. A synchronous and immersive virtual world is the basis for the composition of the metaverse, which also means that with the support of AI, AR/VR and other technologies, everything existing and occurring in real life, including human images, can be synchronized in the virtual world.

The second is "open source and creativity". The open source we are talking about here is not only technology but also the construction of a set of open source platforms. Metaverse will encapsulate and modularize the code to varying degrees by setting "standards" and "protocols" to allow users with different needs to create and form an original virtual world in the metaverse, thus constantly expanding the boundary of the metaverse.

Then, the third is "eternity". The metaverse platform will not "pause" or "end" but will run in an open source manner and go indefinitely.

The last one is the "closed-loop economic system". Users' production and work activities will be recognized as a unified currency that players can use to consume the content on platforms or exchange a certain proportion of real currency. The economic system is the engine that promotes the progress and development of the metaverse.

2021 can be called the first year of "metaversion", and a large number of companies at home and abroad have swarmed into the metaverse track. However, if we return to the technology itself, there are still many technical difficulties and application challenges to be solved in AR and other technologies. According to the statistics of VRPC industry analyses and experience optimization platforms, the main pain points of current user experience are as follows:

1. Lack of high-quality content;
2. High-performance terminals have certain price thresholds;
3. The visual quality is limited owing to the limits of reality technology, resolution and visual field;
4. Head-mounted display equipment causes head movement response delay (MTP), vergence accommodation conflict (VAC) and vertigo;
5. Lack of cloud network optimization for virtual reality services as well as poor network perception;

Now, various types of products related to the concept of metaverse are still far from the complete ideal form of metaverse in the strict sense. In the short term, we should not interrupt the original business plan to blindly pursue hot spots. It is still too early to build a fully immersive virtual world for hundreds of millions of users before AR/VR becomes the next generation general computing platform. At the moment, the top priority of the major hardware manufacturers engaged in AR/VR is to promote the delivery of customer-side equipment by virtue of high-quality content and more competitive prices to pave the way for the subsequent development layout of the metaverse concept. Once technologies mature and content ecology is formed, the grand vision of the metaverse will eventually come true.

7.2 New Risks of Deep Synthesis Technology

If the ubiquitous face recognition in our life is to realize identity recognition through facial feature detection, then deep synthesis is to synthesize virtual images based on the features of objects, including face and human body, and authentically simulate the features of the referent as much as possible. In a sense, deep synthesis is an important approach to build a digital twin world, but it can be seen as a "Trojan horse" to break the exclusivity of face recognition.

Deep synthesis also creates a new research proposition for the further optimization of face recognition—the leap of deep synthesis technology and the continuous rise of its applications, which not only poses a higher challenge to the current biometric recognition technology, including face recognition but also further expands the development prospect of AI application in the field of biological information.

As an AI-generated media technology, deep synthesis was first mainly used in digital special effects in film production, but it came to public attention when someone used it to replace the heroines' face in American adult films with famous Hollywood actresses in 2017. For this reason, the technology was pushed to the cusp of public opinion and has been monitored and restricted by numerous government network regulators. However, in reality, AI deepfake is only a small part of deep synthesis applications. Deep synthesis includes face reproduction, face generation, speech synthesis, etc., and is expected to develop in the direction of full-body synthesis and digital virtual human.

Among them, "AI deepfake", which first entered the public domain, is face swap. As a widely used form of depth synthesis at present, face swaps are mainly used to replace the face of the target task with the face image of a specific character. Typical applications include ZAO and FakeApp. They can realize the most basic face deep synthesis but are not able to achieve high-precision deep synthesis information.

"Face reproduction" mainly involves driving the facial expression of the target character, including the sideways mouth, eyebrows, eyes and head of the target face, to further control the facial expression of the target object. Different from face swap, face reproduction is not to replace the target identity but to change the facial expression of a real person and let him/her say what he/she has never said through simulation.

"Face synthesis" is designed to create a new face virtual image comparable to the real face and can even replace some real portraits for advertising, user account pictures, etc. "Speech synthesis" is used to create a specific voice model to transform text information into tones that are close to natural speech and rhythm.

The technology "GAN" behind deep synthesis has received extensive attention from the industry, and its technical maturity and social influence have been significantly improved since 2017. At the same time, deep synthesis technology has also begun to develop from the early simulation of a single target object in a single audio and video to a comprehensive target object of complex audio and videos. For example, in 2019, researchers from Stanford University, Max Planck Institute for Informatics, Princeton University and Adobe Research Institute realized the technical effect of changing the narrative mouth of the target character in real time according to the input text through the integration of speech recognition, lip search, face recognition and reconstruction and speech synthesis, making the modification of target tasks more natural in video post production. At this point, Carnegie Mellon University matched the lip movement of the deeply synthesized virtual character with the given audio, realized the recognition of multiagent voice features in the same media, achieved many-to-many audio and video conversion, and processed the voice changes of different identity characters simultaneously, further improving the comprehensive video processing capability.

All these above are for the "two-dimensional" depth synthesis of audio and video, and the most promising direction of this technology rests with the virtual digital human formed by three-dimensional depth synthesis. In essence, traditional virtual digital images are based on real characters, which can reproduce the behaviors of real characters through action simulation. Weta Digital, a film special effects company that has achieved a new height in film action capture technology and produced numerous top visual images in films such as *Avatar* and *Rise of the Planet of the Apes*, has tried the first "human like" film role in its work *Alita: Battle Angel,* which not only realized the real-time feedback of expressions and actions but also reached the peak in detail. With Iris, which has over 8.3 million polygons, the film has also realized the combination of virtual human and physical particle effects that are very difficult in digital special effects such as light, wind, rain and water and has set a milestone for the role of digital special effects virtual film once again.

In fact, the commercial application of deep synthesis technology in film and television has been common for many years. With the current popularization of deep synthesis technology, entertainment, social networking and education have gradually begun to use deep learning to optimize their product functions and user experience. The film industry's exploration of virtual synthesis and character synthesis stands at the forefront of all industries. As early as 1989, *Abyss*, a film, adopted expression capture for the first time in human history. James Cameron led the creative team of the film to make the flowing water column with the expressions of human faces through digital special effect technology based on optical expression capture. In just a few seconds, these pictures have made this film a pioneer in human digital special effects. Then, starting from the first motion capture film, *Total Recall,* in human history, virtual synthesis based on motion capture technology has gradually become the basis for the presentation of the magic effects of light and shadow that human beings are unable to reach.

Now, whether the deep synthesis combined with motion capture technology in *Fast&Furious 7* brings Paul Walker who died tragically during the filming back to life on the big screen or creating completely realistic-looking virtual film images based on deep synthesis such as *Rise of the Planet of the Apes* and *Alita: Battle Angel* has brought great innovation to the current film industry. Deep synthesis not only improves the postpresentation effects of audio and video creation and realizes excellent audio-visual effects that are hard to present through traditional shooting methods but also reduces the creative limits of creators and the work pressure of the postprocessing team caused by the restrictions of shooting sites and conditions. Meanwhile, the virtual role further weakens the impacts of actors on the film and breaks through the restrictions of actors themselves on the role. Moreover, it can also flexibly change actors' expressions, actions and lines according to the script in the later stage of film to expand creative space.

Culture and entertainment have positively embraced deep learning. Video synthesis applications, including FaceApp, ZAO and Snapchat, have set off a wave at home and abroad, greatly enriching the playability of image and video postprocessing software. For example, FaceApp can enable users to see their old appearance based on in-depth learning or synthesize their infant looks. Some game manufacturers also set down to try to create user virtual characters in the game scenes based on deep learning technology so that players can incarnate the roles in the games and further increase the sense of presence and participation.

Otherwise, deep synthesis also has a brilliant future in education, social networking, art and digital marketing. In the field of digital marketing, virtual models based on face swaps, face synthesis and other technologies can allow customers to intuitively see how clothing looks on the human body. Beyond that, the virtual images in advertisements can realize real-time interaction with consumers watching advertisements and further strengthen consumers' perception and experience of digital marketing content. Deep synthesis can not only be applied on the consumer side but also needs to be explored in medical treatment, scientific research and so on. For example, in the medical field, AI synthetic virtual characters can help patients with ALS communicate more naturally with others or leave permanent voice copies for patients who are about to lose their ability to speak. In the area of scientific research, the virtual environment and data sets created based on deep synthesis have been applied to the development of application cases such as automatic driving simulation systems and AI assistants to effectively reduce the current difficulties in obtaining data from reality.

Although the advances in deep synthesis in recent years are clear to all, and it has made great achievements in forward-looking application scenes such as film and television special effects and digital twins, a series of biological information forgery problems triggered by deep synthesis are gradually emerging. Making pornographic videos using AI face swaps, disseminating harmful information through synthetic fake politician videos and other such conduct have aroused the industry's concern about the evil side of the technology. Particularly, some people even use these fake digital media content to carry out fraud, identity counterfeiting, and other illegal activities, making information security and network attack and defense face new challenges.

Under the wave of informatization, intelligence and digitization, the iteration of the algorithm model is inseparable from the guarantee of large-scale basic data sets. Different from the textual data used in traditional machine learning, deep synthesis technology is based on a large amount of image data. As the most widely used citizenship identification, data directly determine the right of citizens to participate in public activities in the digital governance environment. Once data are tampered with or used by outlaws, they will not only infringe on citizens' privacy and portrait rights but also intervene or even deprive citizens of their rights to participate in social public activities or directly tamper with a person's identity and social role, resulting in significant risks of leakage and tampering as well as serious consequences. This also puts forward higher technical and moral requirements for privacy protection, network and data encryption, security attack and defense, standard and specification establishment, etc. Apart from that, face forgery based on deep synthesis continues to challenge the security authentication mechanism of face recognition, and the systemic risks of face recognition in the future should be taken seriously.

As face data are the unique identity for each person, the right to use them, namely, the so-called "portrait right", should belong to the natural person ontology. As a part of the right of personality, portrait rights include citizens having the right to own their own portraits, having the exclusive right to make and use their portraits, having the right to prohibit others from illegally using them or damaging and defiling them, or using their exclusive portraits for profit without the consent of the portrait owner. For a series of applications represented by the face recognition combined with deep synthesis of ZAO, their infringements on the legitimate rights and interests of users in user agreements should not be ignored. For example, Article 6 of the first edition of the ZAO user agreement stipulates that "ZAO has the rights to modify and edit users' content and such rights are completely free, irrevocable, permanent, sublicensable. ZAO also has the rights to disseminate the such content on the Internet before and after modification and use all the copyright property rights and neighboring rights enjoyed by the copyright owner under the *Copyright Law*". The behavior of over-grabbing user authorization and privacy data by virtue of technologies has further raised public doubts on the security and reliability of ZAO and the applications of face recognition in sensitive fields such as security, finance, identity recognition and authority management. Taking deep synthesis based on face data as an example, the behavior of synthesizing a person's portraits for commercial purposes without his/her permission can bring great challenges to the existing civil rights protection and judicial protection mechanism. The vague definition and wide impacts of infringements will directly threaten the legitimate rights and interests of citizens.

With the all-around boom of deep synthesis technology, virtual digital images and digital twins are becoming increasingly popular and even become a part of public infrastructure. However, the extensive promotion of deep synthesis technology is actually based on an unequal premise; that is, citizens are the beneficiaries and advocates of the current digital infrastructure dividends. However, based on the data of CNNIC, over 500 million people in China have no access to the Internet at the moment, and a large part of them are vulnerable groups, such as the elderly and the disabled. They may feel inadequate when faced with the product of the high combination of deep synthesis technology and social governance. Thus, they are deprived of the rights to enjoy all social public services and security, and the digital divide further aggravates the differentiation of rights. Otherwise, with the expansion of technology, technology risks are expanding while technology welfare is gradually transformed into basic rules. Since not all people are supporters of technology, everyone has the right to decide whether their biological information and other private data can be used for other purposes, which makes those who want to reduce technical risks, protect their own private data and refuse to use deep synthesis technology lose the right to enjoy basic social public services and infrastructure.

In conclusion, deep synthesis technology uses human behaviors and biological data to create virtual images, which is actually a kind of alienation for people. It is mainly manifested in information privacy, feature instrumentalization, personality labeling, individual datamation, etc. People's traditional cognition of existing things is being challenged and reshaped during this process.

7.3 Masked Virtual Digital Humans

On June 15, 2021, the Department of Computer Science and Technology of Tsinghua University held a press conference on the achievements of "Hua Zhibing", announcing that she was officially "enrolled" in this university. The virtual image was said to possess continuous learning ability and could gradually "grow up". She could learn the patterns implied in data, including text, vision, images, videos, etc., just as human beings can keep learning behavior patterns from the things they experience around them. As time goes by, Hua Zhibing will organically integrate new abilities she learns for new scenes into her own model to increase wisdom.

More companies, including Microsoft, Baidu and Tencent, have launched virtual digital human products, which give realistic portrayals of the pictures in many movie scenes.

The term "virtual digital human" stemmed from the "Visible Human Project" initiated by the United States National Library of Medicine (NLM) in 1989. The "virtual digital human" here mainly refers to the visualization of human structure and the display of the size, shape, position and spatial relationship of human anatomical organs in three-dimensional form. In other words, it is used to realize the digitization of human anatomy. The research results are mainly used in human anatomy teaching and clinical treatment in the field of medicine.

Different from the virtual digital human in the medical field above, the virtual digital human we are analyzing here refers to the human avatar with digital appearance formed by AI technologies such as dynamic three-dimensional reconstruction, CG combined with motion capture, simulated manikin, cartoon modeling combined with speech synthesis, etc. "Virtual digital human" should have the following three characteristics: first, it has human appearance features, specific appearance, gender and personality characteristics; second, it has human-like behaviors and the ability to express with languages, facial expressions and body movements; third, it has a human-like mind and the ability to identify the external environment and interact with people.

From the earliest manual drawing to present computer graphics and AI synthesis, virtual digital humans have roughly gone through four stages: germination, exploration, initial development, and growth.

In the 1980s, people began to try to introduce virtual characters into the real world, and virtual digital humans entered the germination stage. In 1982, after the Japanese cartoon *The Super Dimension Fortress Macross* was broadcast, the producer made Akemi, the heroine, a singer who sang the cartoon collection and produced a music album, and successfully entered the famous Japanese music list Oricon at that time. Akemi became the first virtual singer to write songs in the world.

In 1984, George Stone created a virtual character called Max Heidrum. Max has a humanized appearance and expressions and wears a suit and sunglasses. He had starred in a film and shot several commercial advertisements. He once became a well-known virtual actor in the UK. By then, the virtual image was acted upon by a real actor through special effect makeup and hand painting due to the limits of technologies.

At the beginning of the 21st century, traditional hand drawing was gradually replaced by computer graphics (CG), motion capture and other technologies, and virtual digital humans entered the exploration stage. In 2001, the character Gollum in *The Lord of the Rings* was generated by CG technology and motion capture technology. Then, in 2007, Japan produced the first widely recognized virtual digital character "Hatsune Miku", a two-dimensional style male idol. Her early character image was mainly synthesized by CG technology, and her voice was synthesized by Yamaha's VOCALOID1 series, but her presentation form is still slightly rough.

In the past five years, due to the breakthrough of deep learning algorithms, the production process of digital humans has been effectively simplified, and virtual digital humans have begun to step on the right track and enter the initial stage. In 2018, Xinhua News Agency and Sogou jointly unveiled an "AI news anchor". After users input a news text, the image of the digitised reporter will be played on the screen, and then it will read out the text. Its lip movement can be synchronized with the broadcast voice in real time.

From the analyses of industrial ecology, the industrial chain of virtual digital humans can also be divided into a foundation layer, platform layer and application layer from top to bottom.

The basic layer provides basic hardware and software support for virtual digital humans. The hardware includes display equipment, optics, sensors, chips, etc., while the basic software includes modeling software and a rendering engine. Display equipment is the carrier of digital humans, including 2D display equipment such as mobile phones, TVs, projections and LED displays, as well as 3D display equipment such as autostereoscopic displays, ARs and VRs. Optical devices are used to produce visual sensors and user displays. Sensors can collect the original data of the digital human body and user data. Chips are used for sensor data preprocessing, digital mannequin drawing and AI computing. Modeling software can be used for the three-dimensional modeling of the human body and clothing of virtual digital humans. Rendering engines can render lights, hair, clothing, etc. Generally, the bottom manufacturers have worked in the industry for many years and have formed solid technical barriers.

The platform layer contains a production technology services platform and AI capability platform providing technical support for the production and development of virtual digital humans. The modeling system and motion capture system obtain assorted information of real person/physical objects through hardware such as sensors and optical devices upstream of the industrial chain and use software algorithms to realize human modeling and motion reproduction. The rendering platform is used for cloud rendering of models. The solution platform provides customers with digital humanized solutions based on its own technical capabilities. The AI capability platform provides computer vision, intelligent voice and natural language processing technology capability. The platform layer brings together more enterprises, such as Tencent, Baidu, Sogou and Xmov.

The application layer refers to virtual digital human technology in combination with actual application scenarios, which can be divided into various types to form industrial application solutions. According to different application scenarios or industries, entertainment digital humans (such as virtual hosts and virtual idols), educational digital humans (virtual teachers), auxiliary digital humans (virtual customer services, virtual tour guides, intelligent assistants, etc.), film and television digital human (doubles or virtual actors) have been around. Different forms and functions of virtual digital humans give power to the fields of film and television, media, game, finance, culture and tourism and provide personalized services for users according to their needs.

An ethical issue that needs special attention here is "virtual eternity" based on digital humans. The so-called "virtual eternity" refers to "a theory that human spiritual self can be uploaded to non biological media in the first person to perpetuate the spirit" (*MIT Technology Review* 2019). In other words, virtual eternity is a kind of technology that can make the spirit alive forever. It realizes the digitization of human consciousness.

Some international enterprises have realized the interaction between people alive and their dead relatives through virtual digital human technology. People alive can even chat with dead strangers through a computer interface to exchange what happened after they died. Foreign start-up technology companies, including Eternime, HereAfter, Nectome, Intellitar, Hereafter Institute and MIT Media Lab, are committed to the research and sales of virtual eternity. The advent of the technology, on the one hand, can alleviate the pain of people losing their loved ones, let the deceased appear on the computer screen in the form of a virtual human, and realize the interaction between the deceased and their loved ones as before; on the other hand, virtual eternity technology will give rise to many ethical problems, such as immersing real people in the virtual world and staying away from the real self. As we can see in the British drama *Black Mirror*.

Virtual eternity technology makes use of the Internet, AI technology, digital assistant equipment, communication dialog and other means to make a person's voice and appearance live in cyberspace for a long time while maintaining real-time interaction. To realize such a function, three foundations are mainly needed.

(1) A good deal of data that can be divided into simple data and complex data. The basis for distinguishing simple data and complex data is the difficulty of collecting data information and the manifestation of data information. Simple data cover collected conversations, narrations, life scenes, e-mail accounts, locations, education and employment history, sports, identities of family or

(2) Active and passive data collection methods. The so-called active collection means that the data of the virtual eternal company come from independent collection, while passive collection means that the data of the virtual perpetual company mainly come from the data information provided by users themselves. From Eternim company requiring users to provide all their personal data, to Hereafter Institute collecting user information through 3D body scanning and motion capture, from MIT Media Lab collecting all the data generated from users' brains every day by AI technology to Nectome gathering information from human brains through biotechnology, which seems more difficult, it can be seen that how virtual eternity companies collect data and information from simple to complex, from passive to active.

(3) Visualization of the deceased. Visualization means that users are able to see the dead again, albeit only virtual images. However, those images can move and have their own thoughts. They feel closer to people than emotionless photos. The visualization of the dead is also the main reason why users are willing to pay a huge amount of money to join the virtual eternity project. It is worth noting that the virtual human that virtual eternity companies try to provide is no different from the people in real life except that it has no physical body. Although science and technology at the moment cannot give virtual humans a human body, in the near future, "lost relatives" in the state of virtual eternity will have their own ideas, be able to learn independently and generate their own thoughts when interacting with their relatives.

From the above technical implementation process, we can see several technical ethics problems. The following three aspects are highlighted here:

(1) First, data privacy. Data are not only the most vital element of virtual eternity technology but also the key to ethical problems caused by the technology. To avoid the user privacy problems that arise from virtual eternity technology, users must have complete control over their data, but companies such as Google and Facebook may have no willingness to cooperate. Therefore, the "enhanced eternity" project is also provoking people's controversies regarding data collection and control. Data ownership has always been the core of ethical discussion of big data technology. As described above, virtual eternity technology companies can obtain data in two ways. They are either provided by users themselves or obtained by companies. The former way is ethical for these companies to obtain data, but what truly matters is whether there are ethical risks in the use and preservation of data after getting them.

Will virtual eternity technology companies use data according to certain data ethical rules? Will the deceased data be sold to other companies and individuals? Can they ensure data storage security? Do they have enough technologies to prevent hacker from stealing the data of the deceased? All these bear thinking about.

(2) Second, the ethical behaviors of the virtual deceased. Since the virtual human is highly intelligent, it is the same as people in real life except for the physical body. Hence, in the virtual world, "evil" virtual humans may engage in a series of immoral behaviors. For example, does the virtual dead character murder other characters in the virtual world? Will virtual characters produce all sorts of discrimination? Although virtual characters are not real people in real society, they are real in the virtual world. Authenticity will generate moral problems similar to real society in the virtual world. For relatives alive, virtual eternity technology gives their relatives a second life. However, as AI continues to advance, relatives in the virtual world may be hurt by others, and they may be unable to reproduce again as a consequence. This may be another blow to their relatives in the real world and will provoke social criticism and opposition against virtual eternity companies.

(3) Virtual humans may destroy the real moral world. The continuous growth and mutation of virtual immortal robots will impact the social or moral world. For instance, with the constant enhancement and updating of chat robots, the images of people they depict will change over time. Even within the five years after the users' death, their registered chat robots are likely to have developed into a "more complex stuff".

It can be seen that digital humans are ethically "masked". On the one hand, they give people realistic feelings and make up for the regret of losing their loved ones; on the other hand, they exert a strong influence on the traditional society. Looking at the development of virtual eternity technology from the angle of scientific ethics, we must reflect on the problems it may bring in the future.

For example, will a highly intelligent virtual eternity system create a second virtual world outside the human world? Will the virtual dead in the state of virtual eternity produce their own consciousness and lead to war or economic crises in the real world? Will the virtual deceased get rid of human control and successfully pass the Turing test and be capable of controlling human beings? Will the technology lead to ethical dilemmas in real society and stipulate the ethical norms of real humans and society with their moral norms? These are all worthy of serious consideration.

The above is our discussion on the development of technologies and ethical risks of augmented reality and deep synthesis, which is also the last topic of industrial ethics discussed in the second part. The ethical problems in the virtual reality world have opened us a new door to understand the interactive relationship between humans and computers. How to construct a new type of scientific and technological ethical relationship will become the common proposition of all technology enterprises and our common proposition to enter the intelligent society. What we can do is now to better promote the development of technology and prepare for the worst situation to prevent possible crises, rather than simply take the advances of new technologies and new industries as the guarantees of social progress.

Chapter 8
Start of the "Age of Exploration" of AI Governance

Two hundred years have passed since Charles Babbage, a British computer pioneer, conceived and designed the first fully programmable computer in 1820. However, the core of the ethical challenges brought by the development of information technology remains unchanged; that is, the progress of ethics cannot catch the speedy growth of IT, which leads to a host of conflicts and problems in the ethics of science and technology.

From the perspective of the development of human civilization, people and technology often go together. Each generation is seeking a way to live in harmony with technology, thus shaping different living spaces. When facing various choices in dealing with technologies, humans have adhered to a basic value principle: **no matter how technological innovation accelerates social change, we should still guide the direction of technological evolution based on human nature rather than succumb to the logic of technological change**.

Marshall McLuhan once said, "We shape our tools, and in turn, our tools shape us". There is no doubt that the emergence of AI technology is creating a utopian era full of uncertainties, and what we need to do is to govern and regulate this era to ensure that this future planning based on engineering ideas will not affect the long-term growth of human civilization. Europe has been a "navigator" in the discussion on this topic for a long time.

However, since the debt crisis, the EU has faced a series of challenges, including the refugee crisis, large-scale terrorist attacks, brexit, etc., which have deeply disturbed the internal order of Europe. What reflects these phenomena are the contradiction of the economic model in the European region, the serious lack of risk resilience of the EU as a whole, and the trust crises among the EU countries. Currently, the EU is facing deep-seated challenges, and its progress in high-tech industries such as AI lags behind that of the U.S. and China. Does this mean that the technological sovereignty of the EU is directly threatened and that it is no longer an example of prosperity and stability on the world stage?

Based on the above realities, the EU believes that promoting the progress of industrial policies may be the key to realizing a "sovereign Europe", which means seeking independence and comparative competence in the digital industry and strategic value chain. The European AI strategy is a big move for the EU to realize digitization and has also become a model of global AI ethics research. On February 19, 2020, the European Commission issued a new digital strategy, another programmatic strategy for the digital transformation of the EU after the Digital Single Market Strategy was launched in May 2015. A *"WHITE PAPER On Artificial Intelligence"* and *"A European Strategy for Data"* were also released together with this initiative.

The introduction of the EU digital strategy is not only intended to enhance its technical sovereignty, industrial leadership and economic competitiveness but also expected to set standards for global digital regulation through EU legislation such as the GDPR, in hopes of bringing a sustainable impact on the development of the global digital economy. The initiative is considered to be the start of the "Age of Exploration" of the digital economy.

This chapter is digging a little deeper into the development venation of EU AI ethics and discussing the core demands of AI ethics and governance in the smart era, especially the basic paradigm of AI ethics based on technology governance. Finally, we will discuss the concept facing the era of man-machine symbiosis. the implementation of "anthropocentrism" and the expansion of humanistic value. Before clarifying the logic behind a range of EU AI ethical policies, we first need to understand the construction process of the European technology governance system.

8.1 New Channel or Old Ticket

The power of technology is reorganizing the human order. With the marriage of technology and science since modern times, increasingly powerful technology has been intervening in human society, changing social relations and even the world political and economic landscapes. Engels emphasized the importance of technology to humans themselves as well as the social relations in which they exist when expounding the significance of labor in the process of human birth. Such importance is reflected in the technicalization of mankind itself at present, which is mainly manifested in two aspects. On the one hand, it manifests itself in the "technicalization of the human body", including changing the biological definition of human beings and creating a "super human" through modern technologies such as gene editing and organ transplantation; on the other hand, it also shows the "technicalization of human survival" on a broader level, since all aspects of human survival are profoundly affected by technologies.

How did it all start? By tracing the history of IT and data technology, we can see that human society has experienced the process of mutual construction and influence between technology and society in the past hundreds of years. One of the most impressive scenes is on March 10, 1876. Alexander Graham Bell, an American inventor, recorded the course of the world's first telephone communication-I shouted into the microphone, "Mr. Watson, come here, I want to see you."

8.1 New Channel or Old Ticket

In this historical event, the key spirit and vital metaphor of modern civilization are displayed. Human beings have chosen to see a "new world" that is essentially an "info sphere" created by themselves. Whether we are facing digital network or virtual reality space, it is just the change and extension of "virtual information space". Our current life comprises the real world and "virtual information space". Many value conflicts and ethical choices faced by us come from this key reality.

AI technology is rapidly changing the lifestyle of most of the world's population. Natural human civilization is transitioning into technological "humanoid civilization". To some extent, we are actually on the eve of cyborgized survival. Driven by the power of technology, the original ethics of human and human, human and nature have to incorporate technological dimensions. As the philosophy of technology and responsibility ethics first came into being in Europe, the development track of ethical science there has become a critical epitome of the development of human ethics. Clarifying the development venation of European philosophy of technology is of profound significance for us to understand the underlying logic and deductive changes of the AI ethical framework in today's world.

In the late 1970s, the dual effects of high technologies were notable. German technological philosopher Friedrich Rapp et al. began to explore the consequences of the progress and social impacts of high technology. They kept asking what was progress, the prospect of technological development and the responsibility of engineers. As the ethical problems of modern technology had objectively broken through the scope that technology itself and traditional individual ethics can solve, there was an urgent need for a new technical ethics theory that can let mankind eliminate the plights of the current value conflicts. By then, the "ship" of technology of philosophy had sailed to the field of ethics.

In regard to the evolution of the philosophy of technology in the ethical direction, Hans Jonas, a student of Martin Heidegger and a famous German philosopher, is supposed to be the first person to pop into our minds. He put forward the theory of "responsibility ethics" of technology in *The Imperative Responsibility: In search of an Ethics for the Technological Age* published in 1979. The author systematically discussed responsibility issues from the angle of ontology and answered three traditional problems yet challenged by the new era, including "who do these responsibilities work for?" "where should these responsibilities be applied?" and "who should take the responsibilities caused by technologies?".

Since the 1980s, responsibilities in technological behaviors, prediction of technological consequences, risk research, and the relationship between technology and cultural inheritance have become the focuses of German philosopher Hans Lank and other scholars. He further enriched the thought of responsibility ethics. In 1982, he published the *Social Philosophy of Technology,* which further pointed out that "there has been a lack of systematic philosophical reflection on technology since the beginning of industrialization". It holds that "only from an interdisciplinary standpoint can we understand the overall phenomenon of technology, explain that technology is something connected with other social lives and cultural traditions, and reveal the systematic relationships involving all influencing factors".

Historically, in Germany, the birthplace of the philosophy of technology, the rise of technical ethics research has an inseparable ideological origin from the reflective and critical tradition of German philosophers. Marx once opposed the concept of "moral neutrality" in his criticism of machinery and industry, and he intended to place the functions of technology in a realistic social background. The dissemination of technological ethics research results has attracted the attention of Germany and even the government of other European countries. That's how various committees and research institutions on technological ethics are introduced.

The Technical Ethics Committee, a combination of technology and engineering ethics under specific historical conditions, alleviates people's concerns about technology to a certain extent. In Germany, there are two top-level technical ethics committees: one is part of the parliament, and the other is affiliated with the Prime Minister's office. The two organizations mainly discuss the problems brought by emerging technologies and provide advice, suggestions and decision-making bases for decision-makers. However, neither of them has decision-making power. In addition, the Association of German Engineers (VDI) also included the relationship between "human beings and technology" in the technical evaluation outline and specifically established a corresponding committee. The articles of association of VID suggest that eight value orientations, including personality development, social quality, comfort, environmental quality, economy, health, technical function and safety, should be adopted to express the complicated relationship between technology and society.

By virtue of the participation of the government and organizations, the German philosophy of technology is witnessing continuous improvement both in theory and practice. It not only theoretically criticizes the ideology of technology, understands the ethical problems of modern technology and further estimates the risks brought by technology but also builds a bridge for the convergence of ideas from philosophers and engineers.

The research on German ethics of technology since the 1990s, on the one hand, has deepened the discussion on the basis, scope, function and system of ethics of technology theoretically. On the other hand, it is increasingly focusing on the ethical defense, risk prediction and safety assessment of high technologies such as stem cell research, gene interventional diagnosis and treatment, cultivation of genetically modified food, intelligent information market, nanotechnology, etc. TThis research not only maintains the speculative tradition of German philosophy but also manifests apparent applicability and openness to the future.

Throughout the development of German philosophy of technology ethics, from the rational criticism of technology tools, the reflection on the essence of technology and the discussion of responsibility attribution, to the methodology and strategic choices to solve the problems of ethics of technology, all these above are consistent with the three stages of the development of global AI ethics discussed earlier—from the discussion of the necessity of ethics, the analyses of ethical values, to today's research on the strategic system of ethics.

In general, this development model has undergone the transition from the discussion of the research object itself, the analyses of the spiritual core that indicate behind things and phenomena, to the overall strategic planning. Does it also mean that AI ethics research has stepped on a "fast lane" with rules and regulations to follow? To put it another way, can Europe's preparation of ethics of technology adapt to the impacts of modern social technology?

Take the ethics committee as an example. The "predecessor" of the ethics committee on AI established by today's technology giants is exactly the VID mentioned above. A dilemma facing Google, Microsoft, IBM, Sage, SAP, Baidu and other technology companies is that although AI ethics risks such as algorithm discrimination exist universally, programmers still lack professional knowledge that enables them to accurately judge ethics, and they are also unable to undertake the responsibilities of making key ethical choices. For example, "whether autonomous vehicles should protect passengers inside the car or pedestrians outside". Hence, the ethics committee with relative independence and members under diversified backgrounds turns out to be more suitable to judge these ethical issues.

In 2018, Microsoft established the "AI and Ethics in Engineering and Research Committee" (AETHER), which is composed of heads of product development, researchers, legal affairs, human resources and other departments. It refers to certain industry standards and considers safety and ethics issues to the greatest extent.

Following the announcement of AI principles in 2018, Google set up the "Advanced Technology External Advisory Council" (ATEAC), aiming to help Google address some of the challenges in its published AI principles to guide its research and the development and usage of its products, such as some disconcerting factors in face recognition and machine learning. Remarkably, the committee established to analyze issues related to AI ethics from multiple aspects was formed by non-Google personnel with multiple backgrounds. However, soon after, in early 2019, Google dissolved it. The reason given by Google is that the qualifications of committee members were questioned by their employees. It was said that their members included the CEO of a military UAV company and the head of a right-wing think tank.

Although the ethics committee is not a new concept, its functions are still questioned by modern society. According to an article written by Daniel Munro, a senior researcher at the Innovation Policy Laboratory at the Munk School of Global Affairs & Public Policy affiliated with the University of Toronto and published in 2020 by The Centre for International Governance Innovation (CIGI), a Canadian think tank, some technology companies claim to be making and preparing to implement codes of ethics and establishing ethics committees to better supervise and manage their work. However, they did so mainly to avoid being criticized by public opinion and regulated by public institutions rather than to deal with moral challenges.

Is ethics committee fulfilling the value carried by human rationality? Many critics have pointed out that most of the existing ethics committees have empty shells. and their main role is serving as the "moral cleaning machine" for high-tech enterprises. Although many enterprises and institutions worldwide have been unveiled to the public in AI ethics circles, it is difficult to say whether their original intentions are to regulate their own behaviors or swear customers of their foresight in the ethical field to pave the way for the compliance of projects.

Thus, it can be seen that the European philosophy of technology has not fully prepared for its ethical development in the face of the rise of AI. Kissinger once published a long article *"How the Enlightenment Ends"* in *The Atlantic*, expounding his views on the development of AI. He held that the current human society is not philosophically and intellectually capable of dealing with the enormous challenges posed by AI. Technological changes such as artificial intelligence are impacting the world order produced in the rational era since the 15th century. Since we cannot fully predict the influences of these new things, they may eventually lead to the result that various machines on which our world depends may be driven by data and algorithms and not bound by ethics or philosophical norms.

Kissinger's concern has a fair point. People generally begin to worry that an algorithm society running in a "black box" will solidify or exacerbate social injustice behind an "unexplainable and incomprehensible" curtain.

In fact, science and technology present a neutral meaning to morality. A necessary practical wisdom is that when the relevant laws cannot effectively regulate the objectives, we should consider laying a foundation for more effective legal regulation through flexible constraints in value and ethics. In other words, in the new channel of science and technology, it is necessary to abandon the "old ticket" way of thinking and carry out a complete value trade-off and ethical construction for the new application scenario of AI technology. Specifically, there are two aspects of work.

On the one hand, refined regulation and management in ethics are implemented surrounding the value demands in various fields of AI ethics (especially the relationship between data privacy and data rights), which is top-down. On the other hand, through the construction of ethical value, enterprises and other relevant participants are driven to make certain commitments and put them into practice according to the responsibilities in their respective fields (trustworthy AI or responsible AI systems, etc.), which is bottom-up. By virtue of these two aspects of work, we can promote the progress of corresponding work under an imperfect legal system and open a healthy and orderly "age of exploration" of AI.

In this section, we review the historical path and some significant development thinking of AI ethics in Europe. What we need to see is that no matter what the development stage of intelligent technologies is in, the ultimate goal proves to be the same, that is, to create a fair and just life acceptable to all. In addition, an array of technical ethics propositions we face need to start from this point, which is also the direction we need to grasp as participants in the era of scientific and technological innovation.

8.2 Excellence, Trust and Reliability

As we exemplified earlier, both Facebook and Google have challenged the public's bottom line on issues such as data privacy, while data-based monitoring and experiments are happening all the time. The behavior itself is not a surprise. In Henry James's short story *"In The Cage"* published in 1898, a young waitress obtains

8.2 Excellence, Trust and Reliability

empirical data by observing the behaviors of her customers. Her working process includes curiosity and her longing for the upper class, hesitation and evasion of the existing life. The young girl has obtained a godlike omniscient machine due to the authority relationship through a fixed third-party perspective, which makes us aware of the need to supervise and regulate the development paradigm of information.

Next, let us sort out the exploration of AI ethics and value principles in Europe, which may provide some ideas. Since the 21st century, the research of European AI ethics has changed from the proposal of "responsible innovation", the formulation of the rules of the global digital economy to the realization of "regulatory innovation".

In the European ethics community, it is generally recognized that benign technological development can be achieved by the guidance and restraint of rules. Ethical governance is the foundation of a stable society, and ethical norms should become the core criteria for managing and controlling social security risks. Whether governments, enterprises or personal life are potentially linked to technical ethics. Once a certain norm is widely popularized in society and becomes a generally followed ethical rule, its restrictive effect can sometimes even exceed that of the law. Ethical norms not only reduce the improper use of technologies and risky R&D but also make the development of relevant industries consistent with the long-term interests of the majority.

Based on this understanding, Europe is at the forefront of the world in the rules of the digital economy and AI ethics. "Responsible innovation" is a concept advocated by European scholars in recent years. It is mainly concerned with the close combination of corporate social responsibility and technological innovation practice. It effectively evaluates and influences all links of technological innovation from an ethical perspective to ensure the sustainability and social acceptability of technological innovation achievements. By adopting the concepts and approaches of responsible innovation to regulate or embed the ethical responsibilities of AI, ethical decisions and actions can be realized to effectively promote the sustainable development of AI.

After 2014, Europe took a series of measures to promote the growth of AI ethics. The European Parliament's Committee on Legal Affairs (JURI) decided to set up a working group to study legal issues related to the development of robots and AI in January 2015. The "Draft Report with Recommendations to the Commission on Civil Law Rules on Robotics" issued by JURI in May 2016 called on the European Commission to assess the influences of AI and formally put forward extensive suggestions on civil legislation on robots in January 2017, proposing the enactment of the "Robot Ethics Charter".

In May 2017, the European Economic and Social Committee (EESC) issued an opinion on AI. It points out the opportunities and challenges it has brought to 11 fields, including ethics, security and privacy, and proposes to establish AI ethics norms and a standard system for AI monitoring and certification. Subsequently, as Ke Jie, known as "the last hope of mankind", was defeated by Alphago in 2017, the "first year of AI", the progress of AI has moved into an unprecedented fast lane.

As people's understanding of AI deepens, Europe focuses more on AI. In 2018, the EU put forward an AI strategy for the first time. To support the implementation of the strategy, the European Commission issued a policy document "*Artificial Intelligent for Europe*" and appointed 52 representatives from academia, industry and folk society to form a team called High-Level Expert Group on AI (AI HELP) in June of the same year.

Although Europe takes a highly anticipated attitude toward AI, it also introduced almost "harsh" protection rules. In May 2018, the EU took the lead in proposing the GDPR, formerly the *Data Protection Act* formulated in 1995, strengthening the absolute control of users over personal data.

GDPR gives the data subject seven data rights: the right to be informed, the right to access, the right to amend, the right to delete (the right to be forgotten), the right to restrict processing (the right to object), the right to carry and the right to refuse. Both enterprises registered in and out of the EU should observe the GDPR as long as their processes of proving products and services involve personal data within the EU. Those who violate the law shall accept a fine ranging from 10 million euros or 2% of the company's global revenue of the previous year (the higher shall prevail) to 20 million euros or 4% of the company's global revenue in the previous year (the higher shall prevail).

In essence, the GDPR reflects the fact that European society is enhancing the privacy security and supervision of AI. As a data protection regulatory framework, the GDPR is also the most perfect and strictest privacy protection regulation at present. According to the data released by DLA Piper, a fine of 114 million euros has been generated according to GDPR in less than two years, among which the largest penalty reaching up to 50 million euros was imposed on Google by France for adopting a "mandatory consent" policy in processing individual user data. Moreover, the data it collected contain vast quantities of users' personal information, which was also used for commercial advertising without their knowledge. For example, Google lacked transparency and sufficient information when sending targeted advertisements to users without valid permission from users.

What should be noted is that the EU is also quite cautious in using face recognition technology. In principle, GDPR prohibits the processing of biometric data for the purpose of identifying a natural person. Because of this, it is clearly stipulated on many occasions in Europe that if there is a face capture camera, the images or videos must be actively or automatically deleted within a specified time (such as within 48 h) and cannot be even properly stored like encrypted storage, let alone illegal use. However, does this mean that face recognition has been "sentenced to death" on the continent?

The *Financial Times* once reported that in a relevant document, the EU hoped to develop a set of "worldwide standards for AI regulation" and set up "clear, predictable and unified rules sufficient to protect individuals", which will go further on the basis of the existing obligations of the GDPR. It can be seen that far from lowering its privacy protection standards, the EU will continue to zoom in and review the ethical risks brought by AI technology.

Apart from creatively promoting the healthy development of data compliance and the digital economy through the legalization of data governance, the EU has been gradually improving the formulation of AI ethical principles. In April 2019, it

8.2 Excellence, Trust and Reliability

successively released two vital *documents—Ethics Guidelines for Trustworthy AI* and *A governance framework for algorithmic accountability and* transparency—which are the concrete implementation of the requirements of "creating an appropriate ethical and legal framework" according to the AI strategy of the EU, providing a reference for the creation of AI ethical rules.

The *Ethics Guidelines for Trustworthy AI* call for "reliable AI", which is more specific and operational with clear evaluation criteria than the ethical standards put forward by other parties. "Trustworthy AI" is a kind of moral reconstruction in the era of AI that mainly has two components. First, we should respect basic rights and applicable regulations, core principles and values to guarantee "ethical purpose". Second, trustworthy AI is characterized by technical robustness and reliability because even with good intentions, unintentional harm can occur due to a poor grasp of technology.

The seven key principles contained in the *Ethics Guidelines for Trustworthy AI* are as follows:

(1) Human role and supervision: AI should not trample on human autonomy. Human beings should not be manipulated or coerced by AI systems, and they should be capable of intervening or monitoring every decision made by software.
(2) Robustness and security of technology: AI should be safe and accurate. It should be extremely reliable rather than vulnerable to the impacts of external attacks (such as adversarial instances).
(3) Privacy and data management: personal data collected by AI systems should be secure, private, inaccessible and hard to steal.
(4) Transparency: the data and algorithms used to create AI systems should be accessible, and the decisions made by software should be "understood and tracked by human beings". In other words, the operator should be able to interpret the decisions made by AI systems.
(5) Diversity, non discrimination and equity: AI services should be accessible to all, regardless of age, gender, race or other features. So do AI systems.
(6) Environmental and social well-being: AI systems should be sustainable (i.e., they should be ecologically responsible) and "promote positive social change".
(7) Accountability: AI systems should be auditable and included in the tip-off scope of enterprises to be protected by existing rules. The possible negative effects of the system should be informed and reported in advance.

In terms of content, these seven key principles underlining spontaneous ethical constraints center more on the respect and universal participation of "human beings" and the pursuit of common well-being and stress the reliable attribute of AI technology itself at the same time. These principles have truly transformed abstract protection theories into tangible guidelines, which has been widely recognized by European society.

We find that the EU has adopted a three-step strategy when formulating AI norms: the first is to raise core requirements for trustworthy AI, followed by a large-scale pilot phase launched to obtain feedback from stakeholders; finally, they will lead an international consensus with European values and speak on international occasions.

The EC assumes that this code of ethics drafted by AI HELP is of great value to the formulation of AI development policies and encourages AI developers, manufacturers, suppliers and other stakeholders to actively meet these key requirements to establish a more favorable social environment for the successful development and application of "trustworthy AI". To ensure that the code of ethics can be carried out in the development and application of AI, the EC has launched targeted pilot work within the EU, and it will strive to promote the international community to reach a broad consensus on "trustworthy AI". The impacts of such consensus can also be seen in my working experience as a participator of the national AI trustworthy general group.

8.3 Human Versus AI: "Victory" of Anthropocentrism

There is such a description in the book "*World Without Mind: The Existential Threat of Big Tech*" written by Franklin Foer, a famous legendary American media professional. This is about how algorithms keep looking for appropriate patterns. It says "they torture data until it makes a clean breast of everything, and as a sufferer being tortured, it will say what interrogators want to hear".

This passage has made us think more deeply about AI ethics. Regardless of how fabulous AI-led production can be, we can see the risks it carries. Facing the concept of automation based on computational engineering thinking, we need to build a complete set of risk control mechanisms to avoid "The Gray Rhino". In the last part of this chapter, let us discuss the differences in ethics and governance paradigms between the EU, China and the United States and provide an idea of algorithmic regulation and technological governance based on anthropocentrism.

For the EU, the proposal of "trustworthy AI" has largely helped shape the responsible international image of the EU and build up the long-term confidence of local consumers in the AI industry. Andrus Ansip, vice president of the EC in charge of the digital single market, pointed out that "The moral dimension of AI is not a luxury or addition. Only when we believe in technology can our society fully benefit from it. While ethical AI is a win–win proposition that can become a competitive advantage of Europe, helping it become a trustworthy and people-oriented AI leader." Figuratively speaking, AI ethics, as a lever, has greatly strengthened the soft power of the AI industry in Europe.

Although most people think it is a good idea for the EU to formulate AI guidelines, the manner in which this was done has been disputed. Matthias Spielkamp, co-founder of "Algorithm Watch", a nonprofit organization, believes that the definition of the concept of "trustworthy AI", the core of the code, is not yet clear, and how to implement the regulation in the future is still unknown. Additionally, Thomas Metzinger, a philosophy professor at the University of Mainz in Germany who participated in the drafting of the guidelines, criticized the EU for not banning the use of AI in weapon development.

It should be noted that these ethical principles provide a detailed framework and make a practical assessment of the measures involved. However, being not truly binding decrees, they will not generate any new legal obligations for relevant institutions and personnel. Moreover, the reason why the Frankenstein paradox, technological singularity or even policy programs for industry and trade were not mentioned here is that they are far from the current AI deployment situations and obviously not suitable for translation into a policy framework, although their concept descriptions seem intuitively attractive.

Additionally, a more serious question is whether the GDPR and ethical standards promulgated by the EU will become the "straitjacket" for the development of the EU AI industry. Insiders worry that these rules will restrict the application direction of new generation information technologies such as big data and AI to a certain extent, highlighting the barriers to technological development. This may have a negative impact on the development and use of AI in Europe to a large extent and put EU companies at a competitive disadvantage compared with their counterparts in North America, Asia and other regions. Overrefined standards will make it difficult for many companies, especially small and medium-sized companies, to drive their business forward under the compliance framework.

Three quarters (74%) of respondents said that data protection requirements are the main obstacles to the development of new technologies—compared with 63% in 2018 and 45% in 2017, according to a survey of Bitkom, Germany's largest trade association. Among the 539 M&A professionals from Europe, Africa and the Middle East surveyed in July 2018, 55% said that their transactions had not been completed for fear of violating GDPR compliance requirements.

Although it may bring a host of negative effects, it is a rational choice for the EU to make ethics efforts, which can be analyzed through the dual challenges facing the EU at present.

From the perspective of internal challenges, most European companies have a longer "reflex arc" toward the digital economy. According to the data, in 2017, only 4% of the global data were stored in the EU, and only 25% of large enterprises and 10% of small and medium-sized enterprises made use of big data analyses. In most member states of the EU, data scientists account for less than 1% of total employment. Although major companies can improve their systems via AI technology, small companies face high barriers, such as a lack of technical personnel, high investment and inestimable economic returns.

In response to the rapid development of AI technology, the EU needs to establish a flexible regulatory framework suitable for future technological development and respect key fundamental principles, including social and institutional principles, such as safeguarding democracy, protecting vulnerable groups (such as children) and privacy, as well as economic principles promoting innovation and competition. Hence, the formulation of AI ethical principles has promoted the sustainable development of high-tech industries in Europe and accelerated the digitization process of the continent to a certain extent.

From the external objective environment, the growth speeds of AI vary in different parts of the world, and some regions boast structural advantages. Taking Silicon

Valley as an example, its unique economic structure can support disruptive innovations with strong business application value. Another example is China, whose regulatory environment is relatively loose in terms of personal privacy and personal data, and public and private investment continue to flow into the field of AI. Although Europe has a solid scientific research foundation, it cannot turn promising inventions into real innovations in the long run. As a result, large global digital companies are in shortage. Europe is also lagging behind the United States and China in patent submission and investment.

Since the EU cannot compete with international leaders in AI hard power, such as China and the U.S., the EU is not able to contend with China and the US in terms of investment or cutting-edge research and application in the field of AI. Therefore, it chose morality as the best option to describe the future of skills. However, can this choice truly meet Europe's expectations? The moral dimension of AI is neither a luxury nor an additional function. Andrus Ansip holds that only by building trust can our society benefit as much as possible from technology. In short, trust is a prerequisite for technological development.

It can be seen that the development of European AI ethics is an indispensable procedure to realize the digital transformation of Europe and deal with technological risks rationally. From a stricter perspective of economics, what AI technology actually causes is not risks but uncertainties. To put it simply, risk is the negative effects of events with a known probability distribution, and uncertainty is the unknown probability distribution. Problems such as human-machine information asymmetry and black box algorithms make the uncertainties in the development of AI particularly remarkable. Human beings cannot predict the specific probability distribution of an event, nor can they make a rational and quantitative analysis on the proposition that "whether AI is to help solve the trickiest problem or threaten their survival". Stephen Hawking once said that "powerful AI will be either the best, or the worst thing, ever to happen to the history of our civilization, and we just haven't realized it yet."

From a spatial perspective, global AI ethics research is mainly divided into two opposing camps, the "fine conservatives" represented by Europe and the "passive liberals" exemplified by the U.S. The European camp follows the principle of prudent supervision. Governments set rules and legal provisions for relevant technologies before they mature. Technologies and relevant industries need to make innovations and progress according to law. While the U.S. camp advocates that whatever is not prohibited by law can be done. The government does not take the initiative to intervene and only promotes legislation when obvious risks are found.

Moreover, compared with the centralized regulation path adopted by the EU, the United States adopts the principle of decentralized governance; that is, the Federation participates in AI standardization. States vary in their legislative orientations and regulation methods toward AI.

On the one hand, a string of documents, including *"Preparing for the Future of Artificial Intelligence"*, *"National Artificial Intelligence Research and Development Strategic Plan, "Artificial Intelligence, Automation, and the Economy"*, etc., have guided the industrial development pattern of AI from multiple dimensions, such as ethics, economy, technology and policy support. Trump signed the *American AI*

Initiative on February 11, 2019, which is an important national strategy related to the growth of American artificial intelligence. It is designed to strengthen the national and economic security of the United States and ensure its R&D competitive edge in AI and other related fields. The document outlines the future directions of the country's AI development from five aspects: investment, the ability to make the data resources of the government public, construction of relevant standards, employment crisis management and formulation of the related international standards. It also puts forward key assistance in the fields of smart medicine and smart cities and clearly expresses the exclusion of cross-border acquisitions of major AI technologies from hostile countries. The executive order mainly requires the U.S. federal government to prioritize the R&D of artificial intelligence and devote more resources and capital to the investment and promotion of AI technology.

However, in the face of the legal regulation of face recognition, the U.S. has not formed a set of centralized and unified legal stipulations at the federal level. Each state formulates policies and bills related to biometrics according to its own situations. For example, in January 2019, San Francisco and Somerville issued a ban on face recognition technology and passed a proposal prohibiting the use of face recognition software in public places. In August of the same year, the Massachusetts legislature voted to forbid the government and police to use face recognition technology. The state stipulates that the evidence-based on face recognition technology has no legal effect on litigation. In contrast, the *"National Biometric Information Privacy Act"* enacted by Illinois in 2008 regulates the types of specific use of biological information rather than strictly restricting the application of biological recognition as the first bill regulating biometric technologies in the United States. Specifically, collecting subjects must inform the information subjects of the content, purpose, retention time, and destruction of the collected biological data and implement it in the form of written authorization. At the same time, the collecting subject also needs to formulate a written privacy policy to clarify the retention and destruction period of biological information and prohibit the sale and disclosure of personal biological information without the consent of the information subject or beyond the legal circumstances.

The fundamental difference between the two sides stems from their cognitive differences of scientific and technological ethics. Different from the "rigorous innovation" of Europe, the attitude of the United States is to supervise AI under the principle of encouraging innovation and taking into account the public good, emphasizing guidance and support rather than strong supervision. For the U.S. side, ethical issues should not become the shackles of scientific and technological development, and ethical constraints are not capable of becoming an effective means to solve the risks brought by new technologies. At the policy and legislative level, their attitudes toward face recognition and other applications prove to be more open and inclusive, which leaves enough development, innovation and application space for American AI technology enterprises to promote the rapid growth of the American AI industry and become pioneers in the field worldwide. However, problems such as insufficient supervision and disorderly competition can hardly be avoided. The cognition has to do with American cultural traditions. Historically, immigrants came to the United

States across the ocean to pursue wealth and development opportunities, and they care little about ethics.

However, Europe holds another cognition. The historical accumulation of European civilization makes Europe relatively cautious about technological progress. Europe has experienced more cycles of civilization ups and downs and is more aware of the impacts of new technologies on human society. Thus, it is particularly vigilant about the possible security impacts brought by emerging technologies.

It is hard to judge which cognition is better, and we will never side with either camp. The Chinese people's understanding of today's world is changing from "international community" to "community with a shared future for mankind".

Compared with the former, the latter is more open, inclusive and cooperative. On the basis of respecting the norms of international relations such as sovereign equality, noninterference in internal affairs and peaceful coexistence, it emphasizes the importance of maintaining international fairness and justice, advocating upholding justice while pushing shared interests, the principle of amity, sincerity, mutual benefit and inclusiveness guiding China's neighborhood diplomacy, a new vision of security featuring common, comprehensive, cooperative and sustainable security, a new model of major–country relations featuring nonconfrontation, nonconflict, mutual respect and win–win cooperation, adhering to the principle of extensive consultation, joint contribution and shared benefits to cooperatively promote the BRI, etc., paving a broader way for China's diplomacy and public diplomacy.

The EU and other organizations and countries also realize that cooperation of the human community is needed in the face of AI. Various multilateral organizations and forums are discussing the governance scheme of AI ethics, including UNESCO, OECD, WTO, ITU, etc. The EU will continue to work with these international organizations and "like-minded" countries and collaborate with global players in the field of AI. However, all these cooperations will be based on the rules and values of the EU, including the sharing of key resources and key data, in the hope of creating a level playing field on a global scale.

On May 22, 2019, OECD member states approved the principles of AI, namely, *"The Principles of Responsible Management of Trustworthy AI"*, which contains several ethical principles, including inclusive growth, sustainable development and well-being, people-oriented value and fairness, transparency and interpretability, robustness, security, safety and accountability.

On June 9, 2019, the G20 approved the people-oriented AI principle, the main content of which comes from the OECD AI principles. This is the first AI principle signed by the governments of multiple countries and is expected to become an international standard. It aims to promote the development of AI with practical and flexible standards and agile and flexible governance methods under the people-oriented development concept and jointly promote the sharing of AI knowledge and the construction of trustworthy AI.

Finally, we also offer three thoughts on the "sustainable development of people and technology" here.

First, people go hand in hand with technology in essence. We need to think about value ethics in the profound internal and interactive network relationship between people and technology. The evolution of machines cannot be separated from human beings, and man-machine relations are not abstract. When studying AI scenarios, we need to consider the actor network of products, tools and services used as the research objects to better understand the risks brought by these technologies.

Second, ethical problems can be dealt with through technological governance. As contemporary society has moved into an era of technological governance, how to improve social operation efficiency through science and technology is one of the most significant propositions. In the field of social governance, the technical principles and methods characterized by rationalization, specialization, digitization and even intelligence have become mainstream. How to regulate ethical issues is the key. This can be done mainly in the following ways: social experiment, planning system, think tank system, engineering city, etc.

Third, we should understand that the core of anthropocentrism is human uniqueness and infinite possibility, which are the root of all innovation and human civilization. The abuse of AI technology and the invasion of human data privacy will gradually lose human self-independence, initiative and creativity. While trying to make full use of intelligent technology, we should also maintain a prudent attitude toward its large-scale use, leaving enough space for the reflection and progress of human civilization.

In summary, truly realizing the harmonious coexistence of AI and mankind is not only a problem at the level of technology application but also a broader and greater goal to be established through a positive vision shared by all mankind. The management and control of AI is a global challenge because technology has no borders and no national boundaries, and it is impossible for one country to work behind closed doors. The only way for human beings to prevent the high technologies they created from hurting and destroying themselves is human self-discipline. To establish effective human self-discipline, we need to change and improve the current market economy model and national governance model of our own.

On the road of pursuing human self-discipline, the EU is striving to dominate the world technology pattern by shaping the AI ethics system. The practice of Europe also provides an empirical reference for other countries when they are exploring the governance framework of AI. There is no doubt that in the upcoming AI society, human beings will still occupy a dominant position relative to machines, but they are to focus more on promoting skill transition and providing support and guarantee for those groups who are more likely to take risks. National public policies are made to improve the quality of human life by establishing a symbiotic relationship between humans and machines and encouraging machines to empower human beings to let AI better expand their capabilities.

Chapter 9
AI Legislation in Computational Society

In today's society, the increasingly popular and diversified applications of artificial intelligence are significantly improving the efficiency of production and the quality of people's lives. However, at the same time, it also leads to a huge ethical and legal problem: the boundary of personal data privacy rights is becoming increasingly blurred, and the violations of personal data privacy rights are getting worse.

From the perspective of modern constitutional law, AI technology represented by deep learning is depriving individuals and communities of their freedom as data subjects and "inducing" people to give up their basic rights through efficiency promotion and life convenience. If human beings do not pay attention to these phenomena and guide them well, then such transformation may make the man-machine relationship significantly adverse to human development—machines will not only replace human behavior and thinking and impair people's human dignity but also degenerate them in the direction of animalization and mechanization. These hazards fundamentally challenge the basic ethics of human society (including the principles of humanity, dignity and justice), weaken and disintegrate the core of the spirit of modern rule of law, and shake the foundation of human civilization. This is the most tragic result of the deterioration of AI ethics that we will see. Therefore, systematically regulating technology development through the formulation of AI laws is receiving much attention from governments all over the world.

From a global perspective, artificial intelligence legislation is still in its infancy. In 2017, Germany revised the original road traffic law and issued the world's first code of ethics for autonomous driving.

On October 20, 2020, the European Parliament adopted three legislative initiatives on how the EU can better regulate AI to promote innovation, ethical standards and technological trust.

The European Parliament assumes that improving legal provisions can promote the commercial development of AI in two ways. On the one hand, the initiative can improve business predictability and promote investment by providing legal certainty for developers and deployers of AI solutions; on the other hand, it can boost the development of the industry by creating a fair regulatory competition environment across the world.

Currently, only a few international or regional institutions in the world have proposed formulating a comprehensive legal framework for AI. Hence, the new legal rules on AI laid down by the European Parliament are of special significance.

Facing the necessity and urgency of AI legislation, the current legal research on AI mainly centers on two aspects: the first is subjective thinking about AI legislation, that is, whether AI should be given the status of legal subjects, and the second is the integrative design of AI and law.

For one thing, if we want to study AI at the legal level, a basic category we have to talk about is "How does AI take responsibility?". The premise to solve this problem is to clarify the legal status of AI and its relationship with human beings. On this basis, we can better grasp the legislative "steering wheel" of AI and raise up to the impacts and challenges that strong AI with self-awareness and independent decision-making capability brings to the existing judicial system, personality rights, human subject status and human legal rights.

In addition, algorithms are the core of AI. In the era of AI, algorithms have become the bottom rule affecting the whole world. We cannot talk about legislation without considering the core position of the algorithm. Instead, we should integrate legislative thinking into the AI algorithm and decomposition procedures and control the trend of AI from its sources.

The in-depth discussion of the above research pushes us to develop solutions to digital governance, social regulation and risk prevention from the perspective of AI legislation. This chapter will focus on the subject thinking and boundary of AI legislation as well as AI algorithm governance. Meanwhile, we will reflect on the ethical boundary of the whole AI legislation and the cognition of AI within the predictable range from the perspective of constitutional law and propose that "human dignity" is the aspect that needs the most concern in the whole AI legislation and governance mechanism, which is also a fundamental discussion in line with the development of human society. If we cannot understand this point, we can never determine the basic logic of the interaction between humans and AI, nor can we see the future direction of human social development.

9.1 Who Do We "Legislate" For?

In 2017, the robot Sofia was granted citizenship by Saudi Arabia, which attracted global attention. The AI system is endowed with legal personality. To some extent, it has opened a new era in which robots and humans coexist. With the gradual penetration of AI technology into the whole industry and its outstanding brilliance in the vertical fields, the continuous iterations of algorithms significantly improve the autonomous decision-making ability of AI. The potential ethical risks behind AI technology are increasingly challenging the existing laws. Especially for the new problems relating to power distribution and responsibility that need to be addressed,

9.1 Who Do We "Legislate" For?

clear requirements at the legislative level are urgently needed. *"Toward a new generation of artificial intelligence in China"* issued by the State Council in 2017 proposed that we should strengthen the research on legal issues related to AI and specify the legal subjects, relevant rights, obligations and responsibilities of AI.

From the perspective of legislative logic, if we want to solve the problem of responsibility attribution when AI causes damage, we should first clarify whether AI should be given the status of legal subject. On this proposition, the legal community has mainly formed the following viewpoints from three schools: "subject theory", "object theory" and a view that regards AI as a special existence between "subject and object".

"Subject theory" faces AI with an inclusive and open mind and believes that AI with "independent decision-making capability has the same rationality and free will as human beings. As it shows thinking processes similar to human beings and meets the necessary conditions for the free will of legal subjects", it should be given subject status.

"Strong AI" in the context of "subject theory" mainly refers to the abilities to make independent decisions, think and act freely without the intervention of human power. That is, a strong AI is a complete intelligent silicon substrate with "self-consciousness". In view of this, strong AI will participate in social activities such as human beings and gradually display its remarkable impacts on human beings. The law should establish the subject status of AI and clarify the scope of its rights and obligations. The "subject theory" is clear-cut in its viewpoint, but on this basis, what kind of legal personality should be given to AI has become the focus of academic discussions. The mainstream views include "fictitious personality theory", "electronic personality theory" and "theory of limited legal personality".

First, "fictitious personality theory" starts with the development of legal subjects, expands the coverage of legal subjects, and focuses on solving the problem of responsibility attribution of AI. The "electronic personality theory" assumes that as the degree of independent decision-making of AI increases, its nature as a simple tool is increasingly short of practical significance. It is of constructive value to identify the legal subject qualification of AI by giving it special electronic personality and fictitious legal personality. This theory has been proven by solid evidence. For example, the European Parliament passed a resolution in 2017, allowing at least the most complex autonomous robot to be established as an electronic person. On the one hand, the "theory of limited legal personality" acknowledges that AI has some legal personality and should be given the status of legal subject to enjoy rights and fulfill obligations. On the other hand, it emphasizes its limited capabilities to bear responsibility and consequences, and it should be regulated by a special legal mode or given instrumental legal personality with priority, technical and alternative characteristics to avoid shaking the existing human-centered legal subject system.

Next, let us look at the identification standard of the legal subject of AI. The primary consideration in evaluating whether AI has the qualification of legal subject is to think from the empowerment criteria of legal personality, that is, whether the concept and essence of "personality" can be nested within the technical principles and application characteristics of AI.

In short, ethical personification, which means the standard of giving AI legal personality lies in rationality. The reason why humans have personality is that they have the features of rationality. Artificial intelligence has a rational essence similar to human beings. In the view of scholars who oppose "Anthropocentrism", we should treat all rational beings as human beings from the perspective of morality. As long as this rational relationship exists, no matter how different it is from Homo sapiens in biology, it is not right to disrespect their human nature or treat it in an inhumane way. If we cannot follow this law, how can we treat all intelligent beings equally?

One thing to point out is that whether AI should be endowed with legal personality is not necessarily related to the correct values that should be followed by AI. Even if AI does not have the status of legal subject, correct ethics and values must be observed by AI. Second, as capabilities of AI continue to expand, the responsibilities that AI needs to undertake are also greater. However, at the same time, the restriction based on legal personality will also limit its responsibilities borne by nonhuman beings. Finally, before confirming that AT has moral ability, we should set up reliable standards to judge its moral ability.

As we mentioned before, one of the vital prerequisites for AI to have legal personality is to possess moral ability. Furthermore, the necessary and insufficient condition for AI to have legal personality includes cognitive ability, capacity of intention and moral ability. Cognitive ability mainly refers to the human-like ability to effectively perceive the objective material world and spiritual world. It contains elements such as memory, curiosity, association, perception, introspection, imagination, intention, self-consciousness, etc., and can be embodied through the process that AI receives the information of the environment and other objects around through sensors and uses and creates this information via self-calculation, which is also the cornerstone of rational construction.

In terms of the mainstream "subject theory", there are mainly the following reflections:

(1) The realistic limitations of the development of AI do not confirm the characteristics of the legal response to the real object.

Since the first proposal of AI in the 1950s, after half a century of development, its development still remains in the stage of weak AI. The current AI technology is based on big data of specific application scenarios propelled by deep learning, big data and computing power. It is a kind of primary AI. It thinks, makes decisions and acts on the basis of a large amount of data. Such static artificial intelligence based on big data belongs to weak AI. Its algorithms may achieve exponential performance improvement in the virtual environment but only achieve linear performance improvement in the real data environment, which has certain limitations.

It can be seen that the so-called "strong AI" with "free will" has become a hot issue among the supporters of "subject theory". However, the following three characteristics of the current development of AI have made us question that scholars holding this view have fallen into a state of "utopianism". In terms of research direction, the research on AI is explicitly oriented to strong AI, which is still an illusory direction that will not significantly promote its values and benefits to mankind. In regard to technical architecture, both the current computing architecture based on semiconductor materials and the algorithm development based on binary logic have difficulty realizing the "independent evolution" of AI. The foundation of the emergence of strong AI is that human beings need to break through the shackles of the existing logic circuits and data structures, which is almost impossible in the short term. In addition, the realization of autonomous AI also lacks sufficient theoretical bases. From the perspective of research principles, both scientific research and industrial development need to strictly abide by ethical and moral principles. Even if strong AI can be achieved, the risks far outweigh the benefits. Thus, this Pandora's box should remain closed.

Additionally, the "subject theory" allows AI to exercise its rights as legal subject, but it is reticent about how AI exercises its rights and demands. The essence of right execution is the dynamic reflection of consciousness and the declaration of the intention of reasonable expectation. The purpose of rights is the different consequences of acts and omissions. It is doubtful whether it is necessary to set rights that cannot satisfy demands. Some scholars take Saudi Arabia's granting of "Sofia" citizenship as theoretical support, but the publicity significance of this case is obviously greater than the objective actual situation of AI based on science and technology. From now to the future, AI is still unable to claim its rights and accurately recognize its own citizenship when faced with infringements of rights and interests.

From the basic principles of legislation, the current judicial structure has no obligation or need to give AI legal personality and subject qualification. Even the identification of a "digital person" in the EU is something more of a statement on the propagandistic and strategic level, which is difficult to implement in the form of actual judicial behavior. In the meantime, the stable bottom logic of the current legislation that is being improved according to the years of development experience of mankind also lacks the theoretical bases and implementation conditions for dealing with the advanced proposition of "giving legal subject qualification to strong AI". Advanced legislation that deviates from the objective facts will not only fail to play the role of preregulation but also alienate the "subject" and "tool" relationship between humans and AI, which obviously runs counter to the original intention of human invention and utilization of AI, causing inestimable consequences and risks.

(2) Criticism on the elements of defining "legal personality" in "subject theory"

According to the discussion on the "subject theory" above, we can see that the basis of identifying the legal subject is to establish legal personality and the ability to correspond to rights. The three parts can be explained interchangeably. Based on the "subject object dichotomy", to discuss the possibility of giving AI legal subject status is actually to discuss whether it is necessary to identify AI with legal personality. Under the framework of China's current legal norms and theories, the legal circle mainly acknowledges two subjects, "natural person" and "legal fiction". Therefore, this proposition can be transformed into comparing the legal personality elements of AI with those of natural persons and legal persons to seek reasonable proof of giving AI legal personality from the aspects of theoretical bases and practical needs.

Does AI have the elements of legal personality? AI is a unanimated body created by humans with the help of intelligent technologies. It has no metabolism of life. First, AI lacks the physiological basis of the human-like "carbon-based" brain. The human brain is the core of human life. Brain death means the loss of legal personality. Neither the "silicon-based" AI program nor the AI robot satisfies the physiological basis of legal personality based on the existing legal basis framework. Second, AI does not have rationality and will. Rationality comprises people's perception of the characteristics and laws of things around them, the understanding and compliance of morality and ethics, and the empathy of emotions. Kant proposed that "man is rational and itself is the purpose", Hegel agrees that "man has purposes because of rationality", and man has the abilities of self-reflection and self-examination due to rationality.

Different from human rationality, the "rationality" of AI is essentially a logical relationship based on an algorithmic model without the "inborn" rationality of human beings. The value of its "rationality" also serves human will and purpose. It is impossible for AI to realize self-reflection and self-examination and break away from the restrictions of human regulation at the algorithm level for critical reflection. Even if AI has the ability of deep learning, this algorithm-based correction is essentially a pattern recognition, rather than possessing natural-born rationality. This "wisdom" is the algorithm simulation representation empowered by human beings.

Will is not a mere logical calculus; it contains two key factors: desire and action. Driven by desire, human beings make ethical value judgments on the realistic propositions they face. People's desire and pursuit for survival, quality of life, identity and personality are driving human beings to create a better living environment or make benefit-tending and harm-avoiding judgments by constantly transforming themselves and nature. However, AI has no desire to proceed from its own and public perspectives, let alone emotional constraints. It is difficult to evaluate the rationality of the decisions AI makes when it encounters propositions such as the "trolley problem". AI only performs probability calculus on the basis of human algorithms, not to mention the possibility of establishing free will without the interventions of human beings.

In brief, no matter how we understand AI, its subjectivity cannot be recognized in the short term. What should be especially noted is that although the subject identification of AI is still pending, it actually affects human subjectivity. Maintaining human subjectivity is the basic principle we need to adhere to when discussing AI legislation.

9.2 "What Makes a Machine Human"?

At the level of AI legislation, apart from exploring whether AI has the status of legal subject, another thing we need to pay attention to is the right of human personality. The right of personality can be summarized as personal independence, personal freedom and personal dignity. Specifically, it includes right of reputation, right of portrait, right of name, right of privacy, right of credit, right of personal freedom and so on.

With the rapid advance of AI in the wave of informatization, intelligence and digitization, the iteration of its algorithm model is inseparable from the guarantee of large-scale basic data sets. Different from the data in individual closed environments used in traditional pattern recognition, the current AI possesses increasingly stronger public attributes. Most AI products need to rely on a large amount of real user data, especially the face and identity data used in face recognition. As the most widely used citizen identity, face and identity data directly determine the rights of citizens to participate in public activities in the digital governance environment. Once the data are tampered with or manipulated by criminals, they will not only infringe on the privacy rights of citizens, such as the right of portrait but also interfere with or even deprive citizens of the rights to participate in social public activities or directly tamper with a person's identity and social role. The risks and consequences of disclosure and falsification of face and identity data are more remarkable, which also puts forward higher technical and moral requirements for the carriers of AI, such as the protection of privacy, network and data encryption, security attack and defense, and the establishment of standards and norms. Moreover, as the application of AI reverse generation based on adversarial generative networks and other technologies is impacting the current security authentication mechanisms, the infringements of AI on human personality rights in the future cannot be ignored.

AI obviously has hidden dangers threatening user privacy and security. Whether deep learning, reinforcement learning or federated learning, the basic logic of AI is inseparable from the vast quantities of applications of data training models in real scenes, namely, the collection, use, sharing and destruction of user data.

Given the interconnections and complexity of the data-based AI industry chain and the basic training data, the production factor, running through the whole industry chain of AI, the uncertain risks of the collection terminal, processing node, storage medium and transmission path of data analysis are high, and users lack the control, supervision and knowledge of the above links. Moreover, the industry also lacks unified standards and codes of conduct to regulate the distribution of rights and responsibilities and behavior compliance of relevant enterprises in the above four links. The use boundary and security protection of this part of private data completely depend on the security technology literacy, ethics and industry self-discipline of face recognition service providers. Users have little control over their own face privacy data. The loss of control over personal privacy data may get users into privacy overdraft and make them resistant to modern AI technology.

In addition to high-risk situations such as leakage and falsification of the data used for the training of the underlying algorithm models of AI and the data collected to support personalized products and services, another challenge to privacy ethics is that strong AI products may break through their design ethics bottom line and let users actively offer or even force them to provide privacy data that are not helpful to the learning process or highly sensitive. Especially for all kinds of robots or companion robots designed based on strong AI, once the progress of AI technology is able to endow this type of educational AI product with independent consciousness and behavior decision-making judgment ability, they may maliciously lure and coerce users to disclose their sensitive data and collect them for other nonpublic activities. How can AI products establish reasonable interactions with human beings under the normative framework and do not infringe on user private information without going beyond their functional authority? We need to take the proposition seriously when facing the rapid development of AI technology.

Take the most common biological recognition in the field of AI as an example. For the proposition of dealing with users' biometric information, enterprises represented by Apple claim that they only store users' fingerprints and face data in local terminal devices and adopt physical encryption to ensure complete desensitization and localized storage of the whole calling procedure of the data. However, more manufacturers take face data as a part of user portraits and spread them online at will. The common network transmission and encryption protocols obviously cannot match the security level and sensitivity of face data. Data storage and transmission processes of face recognition applications may be hijacked, causing serious security risks. At present, data crawlers, network intrusion and data leakage have become normal on the Internet, and suppliers collecting face data may take the initiative or be forced to illegally collect, use and spread user face data without the authorization of users or beyond the scope of user agreements.

For example, once the facial feature information is intercepted by criminals or copied from the attacked local encrypted storage and applied to the security authentication service used by the target user, it is quite easy to unlock the user's own face recognition services and obtain the sensitive permissions (such as financial transactions, face access control, mobile phone unlock code and computer sensitive data access) only owned by the user. Some criminals even make facial data into multimedia content that will traduce user personality with the help of deep synthetic technology or maliciously fabricate false video clips to insult, defame, disparage others. These practices are bound to cause great personality insult and inner trauma to victims as well as extremely bad social impacts. These privacy data can even be used to intervene in national political or military actions. The deeper users rely on face recognition services, the more significant the impact of privacy disclosure on users. The specificity and uniqueness of face recognition also make it difficult for victims to recover and eliminate the losses and adverse effects resulting from face data leakage or illegal falsification and utilization.

It can be seen that the potential privacy risks of AI will not only infringe on the privacy rights and interests of users but also face damage to the rights of portraits, names, and credits caused by privacy leakage. The systemic risks among the situations above should be taken seriously.

In addition to privacy and security issues, AI may infringe on the legitimate rights and interests of users, such as life safety and property security.

Take the medical robot as an example. Cultivating a doctor requires more than ten years of hard research and thousands of hours of clinical experience. During the process, the knowledge structure should be closely linked with the experience system. A medical worker is unable to fulfill his/her professional value even with only one link missed. Throughout the whole medical AI industry chain, from the data analysis and computing architecture of the base layer, the algorithm and platform construction of the technology layer, to the scene development of the application layer, an implemented and commercialized medical AI product may build on tens of millions of data, millions of lines of codes, and tens of thousands of parts from dozens of production lines of over a dozen suppliers. The product involves too many enterprises in different fields and directions. Their enormously variable differences in industry backgrounds, technical resources and product judgment standards undoubtedly cast a layer of anxiety on the safety of AI from the patient point of view. For example, disinfection of surgeons and medical devices is a must for artificial surgery. In addition, surgeries involving robots need to thoroughly disinfect surgical instruments. Considering the precision and durability of a machine, we cannot guarantee that a metal machine can be completely disinfected, nor can we ensure it will not be contaminated by other harmful substances.

Thus, it is difficult to guarantee the safety of patients. For example, in a surgical operation of high difficulty and precision involving a surgical robot, any error will directly lead to the failure of the operation and the death of the patient even under the control of a doctor. When this happens, neither the patient nor the doctor can confirm whether the error is a "direct human error" of the doctor who controls the robot on site or an "indirect human error" caused by the robot manufacturer or algorithm provider, or an accident caused by subjective malicious physical attack and remote interference of external individuals or organizations or interruption of power and data transmission. Sometimes only the black box characteristic of AI algorithms or "machine error" of autonomous decision-making. The safety of patients cannot be effectively guaranteed, and the application of AI in medical treatment, especially in the clinical field, looks tough.

AI will also shake the rights of human beings to fairly participate in social activities and enjoy social resources. The AI industry chain gets many related parties together, such as chips, sensors, algorithms, terminals, industrial applications, solutions, security encryption, network transmission and so on. Technology abuse and unfair competition are unavoidable. Once either party adopts discriminatory strategies or unfair competition measures against users or its counterparts, prejudice and inequality will eventually be transmitted to end users and undermine their fair rights. Furthermore, the hardware, algorithms and data sets of AI all have high technical barriers; thus, it is difficult to break the unfair situation once it forms.

In the medical field, although the original intention of driving AI empowered the medical industry, especially the public health system, is to solve the contradiction between the supply and demand of medical resources, balance the existing medical resources and implement the structural normalization of national health management. However, given the great differences in regional development in China, if we radically promote the penetration of medical AI into the existing public health system before it matures, there will be a series of adverse consequences, such as the high retail price of AI medical products, single product type and unsatisfactory product performance and quality. For all the reasons above, the willingness of consumers in remote areas or middle- and low-income groups to buy AI medical equipment will be greatly reduced.

If things go on like this, consumers will not only lose the right to enjoy public health resources but also undergo more negative diagnosis and treatment experience in the follow-up process of seeking medical advice due to the lack of AI medical related supporting equipment and data closed loop. Another possibility is that consumers have to learn and pay more to enjoy the same condition of medical resources and bear a heavier medical burden as the members of ordinary families.

Finally, the fundamental reason why the algorithm brings great challenges and risks to human personality rights is that the goal setting of algorithms oriented by benefit maximization may ignore the proposition that man is the ultimate goal. The functional realization of an algorithmic decision system depends on the collection and analysis of data representing human attributes. "In the era of algorithms, the materialization of human beings relies on data. Human relations are being transformed into data relations, and data are becoming new norms. In the meantime, data are gradually wearing down the personalities of human cognition, and humans will eventually be ruled by the data they create."

Of course, this view exaggerates the materialization of algorithm technology to humans and ignores human subjective initiative, which seems inevitably extreme. However, it points out an indisputable fact that algorithms will never treat people as human beings. Algorithm systems only have data. They have neither human beings nor value judgments of human dignity. However, the person who designs the algorithm can make value judgments, as does the person who decides to use algorithms to make decisions or assist in making decisions. The occurrence of algorithm discrimination is largely due to designers, and relevant decision makers are indulging the applications of algorithm technology intentionally or unintentionally. They try to "beautify" the role of algorithm technology through the objectivity and scientificity of big data. In algorithmic decision-making systems, marginal groups in the statistical sense are insignificant in big data; thus, real-world discrimination can also be reflected in data.

The above is our discussion of personality rights and some thoughts on AI legislation. What we need to understand is that emphasizing human personality rights is to delimit the boundary for the use of algorithms, and the risks of algorithms are also an important appeal for us to discuss the legislative proposition of AI. In the last section, we will discuss the problems brought by "algorithm discrimination" and fully discuss human dignity, the most fundamental demand of modern civilized national legislation, which is also the fundamental starting point for us to understand the whole AI legislation.

9.3 The Foundation of Civilization Comes from "Dignity"

The rise and deep application of intelligent algorithms have accelerated the pace of the era of AI. When AlphaGo defeated the world go champion, people generated hidden worries in their hearts while cheering for technology. When intelligent algorithms are widely used in automatic decision-making in various fields, how can human subjectivity be guaranteed?

Algorithmic bias is the result of discrimination when intelligent algorithms are used for decision-making or auxiliary decision-making. The principle manifestations of algorithmic bias include age discrimination, gender discrimination, consumption discrimination, employment discrimination, racial discrimination, discrimination against vulnerable groups and so on. When automatic decision-making leads to discriminatory consequences, how can one evaluate it rationally and put forward possible solutions? The array of questions are not groundless doubts and worries because intelligent algorithms have integrated into life. Some scholars just call the current society algorithmic society.

For example, algorithmic prediction models are being widely used in many fields, such as personalized pricing and recommendation, credit scoring, resume screening of job application, police searching for criminal suspects, etc. With the in-depth application of big data and algorithms, algorithms are defining people's credit qualification, capacity, quality of life and many other aspects and then determining whether people can obtain loans, find their favorite jobs, the price range of services and goods they can accept, etc. AAlgorithms may even be used in criminal justice systems to assess the recidivism possibility of a suspect and decide whether he/she can be paroled, and the assessment result can affect sentencing.

In short, under the third wave of AI development represented by machine learning algorithms, the production process of algorithms has undergone essential changes. The change not only means the improvement in the application ability of algorithms and the popularization of its application scope but also means the expansion of the impacts of algorithms on human society and the prominence of the corresponding governance challenges.

In the second chapter of this book, we have discussed the complexity of algorithm rules. When we apply algorithms to different fields of human society, they will inevitably pose many governance challenges, the most representative of which is "algorithm discrimination".

The big data research reports released by the White House in 2014 and 2016 all focused on the phenomenon of algorithm discrimination. In November 2015, the "*Meeting the challenges of big data*" released by the European Data Protection Board (EDPB) also emphasized the problem of discrimination in big data and algorithms. In November 2018, the "*Public Attitudes Toward Computer Algorithms*" survey released by the Pew Research Center also showed that 58% of American respondents believed that computer programs would always reflect a certain degree of artificial bias.

Algorithm discrimination is mainly reflected in the following four aspects. First, the racial discrimination caused by algorithms is more hidden. Tangible racial discrimination can be easily and accurately targeted, while covert racial discrimination is hard to prevent. The "discrimination Trojan horse" hidden in the algorithm embedded with racial discrimination codes is easier to flourish under the high-tech packaging of AI featured by "objectivity, impartiality and science" and is hidden under the cover of the black box characteristic of the algorithm.

Take face recognition technology as an example. As the face recognition system is becoming standardized and being gradually applied to schools, gymnasiums, airports, transportation hubs, and especially the police system, new harm to people of color caused by racial discrimination in face recognition technology is becoming increasingly prominent. What should be paid more attention is the fact that algorithms applied to reduce bias exacerbate racial discrimination. For example, Predpol, an application already applied to reduce artificial bias in the police in several states of the United States, is an algorithm that can predict when and where crime may occur. However, in 2016, when the human rights data analysis team applied the Predpol algorithm simulation to drug crime in Oakland, California, it repeatedly sent police personnel to areas with a high proportion of ethnic minorities, regardless of the real crime rate in these areas. In recent years, racial discrimination resulting from algorithms has emerged, which fully illustrates the urgency and significance of anti-racial discrimination in the virtual world.

Second, the gender bias arising from algorithms is essentially the extension of the long-standing real-world concept of gender discrimination in the virtual world. Big data is the product of society. Human unconscious gender discrimination will affect AI algorithms that analyze big data and may inadvertently strengthen gender discrimination in employment, recruitment, university admission and other fields.

"The core of algorithms is to imitate human decision-making... In other words, algorithms are not neutral."

For example, when employers type the word "programmer" into an automatic resumption screening software, its search result will give priority to the resumes from male job seekers because the word is more closely related to men than women. When the search target becomes the "front desk receptionist", resumes from female job seekers will be displayed first. In November 2019, David Heinemeier Hansson questioned why his line of credit given by Apple Card was 20 times that of his wife, who actually had a better credit score than him.

Additionally, age discrimination caused by algorithms is the most difficult form of discrimination in the workplace to prove. In employment recruitment and employee management, all data, such as the name, personality, interest, emotion, age and even skin color of employees, are often collected in full. For instance, for people of different ages looking for new jobs, their daily work may include searching internet work websites and submitting online applications. This seems to be a very transparent and objective process that places all applicants in a level playing field based solely on experience and qualification on the surface, but age discrimination can actually be seen everywhere.

In 2016, the ResumeterPro project team of ACCESSWIRE found that up to 72% of resumes would be rejected by the applicant tracking system before manual review. The process is completed through complex algorithms, which may lead to unconscious discrimination based on inaccurate assumptions. Employers can use these algorithms to specifically eliminate applications according to age. It is disturbing that such blatant discrimination is difficult to detect because it is hard for applicants to prove that the reason behind refusal is age.

Villarreal V. R. J. Reynolds Tobacco Co. provides a textbook case of how algorithms trigger hidden age discrimination. Villarreal repeatedly applied online for work with R.J. Reynolds Tobacco Company but had not received a reply until the scandal that "the company screened online applicants according to their age but did not disclose it to any rejected candidates" was reported.

Moreover, the consumption discrimination of the algorithm "calculation" is difficult to prevent. In the era of algorithms, commercial Internet platforms perform accurate digital portraits and digital filings for consumers by digging into their previous consumption data and browsing records. By doing so, algorithms can gain insight into their preferences and easily make differential pricing for users in different regions and periods of time in the hope of maximizing profits. For different market segments, the same products and services can make more profits through differential pricing. As a normal business strategy, differential pricing is a reasonable pricing behavior for enterprises to achieve profit maximization. It exists in fields where prices are prone to fluctuations, such as air tickets, hotels, movies, e-commerce, and travel. As long as the pricing behavior of enterprises is open and transparent and consumers are willing to accept it, there is no fraud in this process. Nevertheless, differential pricing has a clear boundary that businesses cannot raise prices discriminatively against a specific individual or group. In 2017, the nonprofit organization ProPublica analyzed insurance premiums and expenses in California, Illinois, Texas and Missouri. The results showed that some major insurance companies charged 30% higher fees to minority communities than other regions with similar accident fees. Amazon's shopping recommendation system, online travel website Orbitz, Ctrip and Didi Dachce have all been under suspicion of big data-enabled price discrimination behavior against existing customers.

Algorithmic discrimination is much more serious than many people have realized, and it challenges the fundamental principles of human society, such as equality and dignity, undermining human value. Algorithmic decision-making uses the inductive logic method, which refines historical experience and summarizes certain rules. However, the logic of induction itself may have deviation. It tires raising facts to norms, which challenges the traditional dichotomy of "norm-fact" and contains social system risks, because the algorithmic model acknowledges the existing social reality and denies the transforming effect of norms on reality. The algorithm prediction model is future oriented on the surface, but it is essentially past oriented. The development of human society should not be confined to its past state; otherwise, it will immerse people in the vortex of the past forever.

More importantly, the application of algorithmic decision-making in the field of public power shows that algorithmic power is embedded in the operation of traditional public power, which may lead to the risk of power abuse since technology may be found hand in hand with power. In addition, algorithmic power may make the existing power restriction mechanism fail to a certain extent. During the long-term operation of public power, the "technical rationality" of algorithms covers it with a veil that can skillfully avoid democratic supervision. The intervention of algorithms in the field of public power challenges the traditional principle of exclusive power and the principle of due process; thus, it is difficult to implement effective power control, and the anomic power will eventually lead to the erosion of individual rights. In the emerging power pattern in which technological power coerces public power, people will increasingly lose the ability to resist, and most "technological illiterates" can only become the "the meat on the chopping block" under the dual repression of technology and power.

It can be seen that the rise and alienation of algorithm power have obviously reflected the risks and irrationality of technology. However, there are no specific legal rules regulating algorithms. Allowing the development of algorithm technology is likely to throw human society into a worrisome situation. Although intelligent algorithms have unimaginable computing power and speed, it does not mean that algorithms are beyond the scope of human understanding, nor does it mean that human beings do not have sufficient regulation ability in response to their risks.

In his speech entitled "*der Mensch im Recht*", Gustav Radbruch pointed out that for the style of a legal era, nothing is more important than the perceptions of people, which determines the direction of law. Today's society is gradually entering the era of AI. When intelligent algorithms are defining people and making decisions, human value will be weakened if we do not have a clear understanding of the human image. Therefore, it is necessary to reshape the value connotation of human dignity and standardize the value orientation of algorithm application.

We propose three aspects of thinking based on the value system of human dignity here.

First, safeguarding human dignity has become the consensus of modern civilized society and the value of the constitution. The principle of inviolability of human dignity is not only the starting point of the normative system of constitutional basic rights but also the basic value principle of constitutional power limitation norms the overall constitutional system, thus forming the basic value principles of the overall system of legal norms of a nation.

Second, the core value of human dignity lies in emphasizing that human beings should be treated as an end in themselves, which constitutes the highest value of the constitution. The exposition about human dignity from Kant's philosophy laid the tone of the rule of law order in postwar international society. This is all about how humans understand themselves, how to understand people as a whole and how to understand human relationships. Furthermore, it also involves the relationship between individuals and countries formed by communities.

Third, the core of human dignity is to protect human subject status and respect human diversity. To safeguard human subjective status is not only to stress the abstract meaning of humans or the whole of humanity but also to point to the rights of independent individuals who should be respected. The so-called respect for human diversity means that the value implication of human dignity lies in respecting human diversity, accommodating the uniqueness of each individual and prohibiting discrimination.

In summary, ubiquitous algorithm discrimination is closely related to the embedding of discriminatory factors in the operation process of algorithms. However, what is worth pondering is that the current society is not fully aware of the limitations of algorithmic decision-making logic. Blind technology worship or beneficial temptation leads to the lack of necessary evaluation and sufficient demonstration for the

9.3 The Foundation of Civilization Comes from "Dignity"

use of algorithm decision-making. Some scholars have even questioned whether the so-called "smart court" and "smart city" are borrowed prosperity. Since such a boom is permeated with instrumentalism and pragmatism, it is likely to ignore the necessary reflection on human value. From the perspective of the value of human dignity, algorithmic discrimination is suspected to materialize human beings, weaken human subjectivity, deny the diversity of human and social development, and deviate from the core value, value purpose and value implication of human dignity. Therefore, how to reasonably push the applications of AI in the whole legislation and governance system is what we are working to solve.

Chapter 10
New Rules and New Order in the Era of AI

AI is exerting an increasing influence on human beings, with the economic field being most affected. After all, in the eyes of most people, they don't care whether the singularity will come, what they care more about is whether they will compete with machines for jobs. Posts that need repetitive work may be replaced by machines. However, from a historical point of view, humans have updated technology heaps of times. The final results show that new technologies have created more jobs and that the quality of human life has also been improved as a whole. Facing the wave of AI, the economy and society are undergoing rapid changes. The old social contract established by mankind needs to keep pace with the times to accommodate these new technologies, and our economic model should also change accordingly.

In 2016, an interview of *"Wired"* conducted by its reporter Scott Dadich was attended by former U.S. President Barack Obama and director of MIT Media Lab Joi Ito. In the program, Obama also mentioned that in the age of AI, if we want to complete a smooth transition, the whole society should extensively discuss how to deal with the sharp contradictions generated in this transitional period and how to ensure sustained and inclusive economic growth through effective policies.

Intelligent AI has greatly improved our production efficiency, but if we leave it unattended, most social resources will still flow to a small part of the upper class. Therefore, how can we solve the problem and ensure that everyone can obtain a decent income? How should the social security policy in the era of AI realize a fair and effective allocation of resources? In this chapter, we will focus on how policies adapt to the development of technology and the economy under a new social contract. In addition, then we will talk about the ethical issues relating to algorithm regulation and decision-making.

10.1 Creating a Technological Contract Between Humans and Nature

To understand how a new social contract is reached, the premise is to understand the evolutionary process of social contract theory. Hence, let us start with how social contract thought evolves and the basic logic of its contemporary changes.

The thought of social contract originated in ancient Greece. It has experienced thousands of years of precipitation and accumulation from germination to prosperity and then to revival. As an important paradigm of philosophical theory, social contract theory has inspired research in various disciplines, such as moral philosophy, political philosophy, legal philosophy and even economic research. It is one of the vital basic theories when we discuss issues such as the origin of state, society and norms, the legitimacy of government, social norms and rights and so on.

Modern social contract theorists, such as Hobbes, Pufendorf, Locke and Rousseau, developed social contract theory into a complete theoretical paradigm. The criticism of Hume, Hegelianism, Marxism and sociological positivism on the theory of the social contract has made the thought gradually decline in the 19th century. Until the second half of the 20th century, Rawls once again put the fading model of social contract theory into use, causing the emergence of its renaissance and a large number of social contract theorists, including Gauthier, Buchanan, Scanlon, etc. Rawls' work enlightened the academic circle, and he said that "social contract theory is not only a possible interpretation of the origin and legitimacy of the state but also the most convincing explanation of the origin and legitimacy of all the normative systems we regard as significant in the public world." Gauthier and Rawls constitute two typical orientations in the field of contemporary social contract theory. Their thoughts represent the two basic intents of the normative thought of Western politics and moral philosophy, namely, the so-called "Hobbes orientation" and "Kant orientation".

Therefore, in what ways has the development of AI technology changed the social contract? The issue is also related to the development of social contract theory itself. We can analyze it from three aspects.

First, the process from the combination to the separation of social contracts and natural law theory has been accelerated thanks to the development of AI. Social contract theory and natural law and natural right theory are essentially two different sets of theoretical models. In early history, they were not only related to each other but also had unique development trajectories, each maintaining some stable structural elements, thinking modes and value functions. In terms of the legitimacy of rights, social contract theory holds that legitimacy comes from the contract concluded jointly, while the natural law and natural rights theory believe that the legitimacy of rights lies in the natural law and natural rights norms that meet higher standards of eternal justice. One of the greatest characteristics of modern social contract theory is the creative combination of the two theoretical paradigms, with the natural state, natural rights and natural law being its basic theoretical elements. Among them, the natural rights and the natural law often become the substantive rules to be observed when concluding contracts or the substantive standards to judge the results concluded.

10.1 Creating a Technological Contract Between Humans and Nature

However, the theory of natural law has received many criticisms in modern times. Opponents assume that modern social contract theorists insist that the natural law and natural rights are eternal, but the continuous change of history and experience has verified the variability of rights. Moreover, actually, "nature" takes everything in life coolly and never distinguishes the good from the bad. Nature has no natural legitimacy connotation in itself.

Therefore, contemporary social contract theory no longer relies on the theory of natural rights, and they start to separate themselves from each other. In the theories of contemporary social contract theorists such as Rawls, Gauthier and Scanlon, all of them put more emphasis on the "purity", formality and procedure of social contract, and the more neutral formal rules that the contracting conduct itself should abide by, and then deduced the substantive legitimacy standard through contract consensus while denying that a priori exists in the substantive natural law and natural rights criteria before contract concluding. The development of AI is actually reshaping new contract consensus and new standards for human society to get along with machines. What we need to see is that the natural law in the era of intelligent society is no longer completely legitimate.

Second, the hypothetical position replaces the actual position and becomes the basis of the initial state in the contract of an intelligent society. Is the initial state a state or stage actually existing in history, or is it just the imagination and hypothesis of scholars? Is social contract something truly exists or just a normative presupposition? Modern social contract theorists and contemporary social contract theorists hold different views on this. Modern social contract, an initial state before contract conclusion, is a real development stage in human history and a historical fact that can be described, which became the logical starting point of their theories and was collectively referred to as the "state of nature". For instance, Locke made it clear that we cannot presume that a natural state does not exist because of a lack of records on it. Since we were not able to use words until the dawn of the civilized society, the human society before the invention of words is unknown to us. He believed that the natural state is the overall survival profile of human beings before entering the civilized society. The practical position was fiercely attacked by Hume, Durkheim, et al. in modern times. With the development of natural science, this position is even less acceptable.

Therefore, contemporary social contract theorists have sublated this point and turn to a hypothetical position. As Rawls pointed out, the original position can never be regarded as an actual historical state, let alone the real original state at the beginning of civilization. It should be understood as a pure hypothetical state used to achieve a certain concept of justice. In the era of AI, because human beings and machines do not evolve simultaneously, it is the former that shapes the latter. Thus, human beings regulate and influence the ideas of machines through hypothetical positions, which requires a hypothetical position to replace the actual position. This has become the basis of new contracts. In fact, the justice of human society has become the standard and consensus of justice in the machine age, affecting the later process of civilization.

Third, in the age of AI, we focus more on constitutive rules rather than regulatory ones. Regulatory rules are used to adjust behaviors that exist before the rule and tell people what to do in a certain scene. Constitutive rules adjust the behaviors that need to be adjusted and generate the behaviors they adjust. Its typical form is "X is counted as Y in C". In the view of modern contract theorists, the state of nature is real, and the rules of natural law are not specially formulated for contract behaviors. They exist independently of human interaction behavior. People can find and adjust the existing rules of natural law to conduct true and good contracting behavior. Contemporary contract theorists hold a hypothetical position on the original position, and contracting rules are specially "designed' for contracting behavior. Logically, these constitutive rules have defined the original position, social contract and contracting behavior itself, and the emergence of the latter depends on the construction of the former.

Rights in the era of AI are actually most concerned about constitutive contracting principles; that is, both the adjustment of behavior and the mechanism of how to realize these adjustments are major concerns. As algorithms are the basis of AI technology and algorithm regulation is the way to conclude and realize the contract between human society and machines, it requires us to consider not only how to restrict and guide the behaviors of machines but also how to achieve such contracts and guide such practices through codes, namely, law. Thus, it is necessary to focus on constructive contracts rather than adjusting contracts.

In summary, we have seen the development of contemporary social contract theory and the possible new contract logic and directions of human society in the face of AI technology. If the natural law is the basis of a traditional social contract, the new social contract breaks away from the natural law and realizes the new contract between human and machine with a constitutive institutional logic, which is the foundation of our understanding of AI public policies.

10.2 "New Rules" of Human-Computer Interaction

The rise of AI is pushing fundamental and comprehensive changes in economic, social and political fields. The governance challenges arising on this basis require innovative reconstruction of the social public policy framework. At the risk level, the traditional governance structure and methods have been unable to adapt to the new changes, which constitutes the ethical and governance challenges of the AI age. The existing policies of countries follow the logic of "no license regulation" or "early warning regulation", which is limited to the two-dimensional level. The policy that only chooses between innovation and security seems inevitably partial. Human society urgently needs to establish a flexible public policy framework to lay a solid institutional foundation for the age of AI.

In the past few years, from Japan, Singapore and India in Asia to Canada and the United States in North America and then to European countries, countries worldwide have introduced AI policies according to their national conditions. As the first country in the world to release a national strategy of AI, Canada has been trying to become an international leader in the area of AI research. In contrast, Finland concentrates more on achieving global leadership in the applications of AI technology.

10.2 "New Rules" of Human-Computer Interaction

Since 2017, the focus of China's AI policy has shifted from AI technology to the integration of technology and industry. The intensive introduction of China's AI policies means that AI has risen to national will in the context of global competition. General Secretary Xi Jinping stressed at the ninth group study session of the Political Bureau of the CPC Central Committee that AI is a strategic technology leading this round of technological revolution and industrial transformation, having "The head goose effect" with a strong spill over drive… Accelerating the development of the new generation of AI is an effective strategic method for us to win the initiative in global technological competition, and a significant strategic resource for the leapfrog development of China's technology, industrial optimization and upgrading as well as overall improvement of productivity.

Local governments have also issued relevant policies to escort the growth of AI. For example, in the *"Action Plan for Accelerating the Construction of New Infrastructure (2020–2022)"*, Beijing mentioned that the city is to promote the integrated application of new generation information technology such as AI and high-end equipment such as robots with the industrial Internet and underline the construction of computing power, algorithm and computation of the basic layer of AI. In the *"Guiding Opinions on Accelerating the Integrated Development of Southern Shandong Economic Circle"*, Shandong Province noted that it is necessary to speed up the construction of new infrastructure such as AI and propel collaborative innovation in the fields of AI, equipment manufacturing, biological medicine, etc. What is noticeable is that Chongqing has issued two AI policies, among which it is mentioned that the AI public service platform will implement 22 projects with an investment of approximately 28.4 billion yuan. However, across the country, AI policies are largely concentrated on the technical and industrial levels. Relevant laws, regulations, ethical norms and policies have not yet formed a perfect governance system. AI policies and legislation are still constantly clarifying their paths in innovations to find the optimal solution in "gradient descent".

In the *"Development Plan of the New Generation Artificial Intelligence"* issued by the State Council in July 2017, China made a basic design for the ethical development of AI. The conception is divided into a "three-step" strategy.

Step 1: By 2020, the overall technology and application of AI will synchronize with the leading position in the world. The AI industry will become a new important economic growth point. A number of globally leading AI backbone enterprises will be cultivated, and the core industrial scale of AI will exceed 150 billion yuan. This would drive the related industrial scale to exceed on trillion yuan. The first step concentrates on developing core technologies and industries and strengthening the competitiveness of China's AI industry. It does not place too much stress on AI ethics.

Step 2: By 2025, the fundamental theory of AI will make great breakthroughs, and some technologies and applications will reach an internationally advanced level. The core industrial scale of AI will exceed 400 billion yuan, which would drive the related industrial scale to exceed five trillion yuan. Laws and regulations, codes of ethics and the political systems of AI will be preliminarily established. This step clearly puts forward the systematic construction of AI ethics, but it still remains in the phase of "initial establishment".

Step 3: By 2030, the theory, technology and application of AI in China will generally reach the world leading level, and China will become the world's main AI innovation center. The core industrial scale of AI will exceed one trillion yuan, which would drive the related industrial scale to exceed ten trillion yuan. A batch of globally leading AI technology innovation and talent training bases will be formed. It is worth noting that this plan is not only a technical or industrial development plan but also includes social construction, system reconstruction, global governance and other aspects, striving to "formulate more complete AI laws, regulation, ethical norms as well as policy system". From this point of view, there are still many gaps in China's AI policies, so there is much room for innovation.

After thinking about China's AI policies and basic thoughts, let us systematically explore the development path of AI in the U.S. On the whole, as a superpower in the field of global technology, the country has accelerated to seize the commanding height of the development of AI and has grasped the strategic initiative of global competition.

On February 11, 2019, the White House issued the *"American AI Initiative"*, an important national strategy related to the development of American AI. It aims to strengthen the national and economic security of the United States and ensure that the country maintains its R&D superiority in the field of AI. The initiative has pointed out the development direction of AI in the U.S. for the time to come from five aspects: investment, the ability to share government data resources, the construction of relevant standards, the response to employment crises and the formulation of relevant international standards. It also puts forward key assistance in fields such as smart medicine and smart cities and clearly expresses its repulsion of cross-border acquisitions of key AI technologies from hostile countries. The executive order mainly demands the U.S. federal government to give priority to the R&D of AI and to devote more resources and capital to the investment and promotion of AI technology.

The United States has formed four major documents on the development of AI (including the previously released *"Preparing for the Future of Artificial Intelligence"*, *"National Artificial Intelligence Research and Development Strategic Plan"* and *"Artificial Intelligence, Automation, and the Economy"*). They guide the development of the industry from multiple dimensions, such as ethics, economy, technology, and policy support. To summarize the overall development pattern of AI in the U.S., we can start from the following three aspects.

First, the country takes the lead in designing the strategic layout of AI and making AI a national priority. In 2016, three reports with global influence were released. In October, the National Science and Technology Council (NSTC) issued the *"National Artificial Intelligence Research and Development Strategic Plan"*, comprehensively arranging and determining to invest in AI for a long time. At the same time, the Oval Office released *"Preparing for the Future of Artificial Intelligence"* to deal with the potential risks brought by AI. In December, the Office released the *"Artificial*

10.2 "New Rules" of Human-Computer Interaction

Intelligence, Automation, and the Economy", which emphasizes the impacts of AI-driven automation on economic development.

Second, the state continues to strengthen strategic guidance and consolidate the global lead in AI. In 2019, U.S. president Trump's signing of the *"Executive Order on Maintaining American Leadership in Artificial Intelligence"* marked the launch of the "The American AI Initiative", which aims to maintain and consolidate the leadership of the United States in the field of AI. In June 2019, NSTC released *"The National Artificial Intelligence Research and Development Strategic Plan: 2019 Update"*, which updated the original seven strategies on the basis of the 2016 version, evaluated and adjusted AI priorities, and added the eighth strategy of "expanding public–private partnership".

Third, the federal agencies in the United States work to promote the development of AI. The Ministry of National Defense issued the *"Summary of the 2018 Department of Defense Artificial Intelligence Strategy-Harnessing AI to Advance Our Security and Prosperity"* and established the Joint Artificial Intelligence Center (JAIC), which was designed to accelerate AI to rapidly enable the key operational missions and coordinate the R&D projects of AI. The National Defense Authorization Act for Fiscal Year 2019 approved the establishment of the National Security Commission on Artificial Intelligence (NSCAI) aiming to comprehensively review and analyze AI technologies and systems. The U.S. Department of Commerce set up the White House Labor Force Committee to help the United States reserve talents needed for the development of emerging technologies such as artificial intelligence. The NSFC continues to support basic research fields of AI, including machine learning, computer vision, robotics, etc. TThe Congressional Research Service and Harvard University released the *"Artificial Intelligence and National Security"*.

Long-term and sustained investment is a key factor for the United States to lead the world in AI. The superpower has always listed AI as the priority of federal investment, emphasized the core role that AI plays in emerging technologies, and concentrated on sustainable investment in basic fields with potential long-term benefits. These measures have provided continuous innovation ability and core competitiveness for the growth of the American AI industry and laid the foundation for America to be the leader in this field.

Generally, China and the United States have many similarities in AI policies in innovation, talent, basic research, R&D and the transformation of scientific and technological achievements. Meanwhile, there are also obvious differences between them. From the perspective of research scope, China's research is more micro and focused more on the present, while the research in the Americas is macro and concentrated more on the future. In terms of research content, China conducts deeper research on fintech and scientific and technological agriculture, while the U.S. offers more support to health care and education. Although the two countries have different policy focuses, their common feature is that they take a generally inclusive and open attitude toward the development of AI and the challenges it raises. Their policy objectives are more inclined to boost technological innovation to maintain the dominant position of national competitiveness. By comparison, countries such as Britain, France and other European countries take different policy paths.

In 2016, "*Artificial Intelligence: Opportunities and Implications for Future Decision-making*" issued by the British government elaborated the revolutionary impacts of AI and planned how to employ it, with particular attention to the legal and ethical risks brought by AI development. In this report, the British government stressed the impacts of the combination of machine learning and personal data on basic rights such as personal freedom and privacy, clarified the concept and mechanism of accountability for decision-making involving AI, and made provisions on specific policies such as algorithm transparency, algorithm consistency and risk distribution. Similar to the UK, France released "*French Intelligence Artificielle*" in 2017. In this document, France continued the legislative spirit of the "*Information Society Act*" passed in 2006 and emphasized strengthening the "common regulation" of new technologies to fully protect individual rights and public interests while enjoying the welfare improvements brought by technological development.

Through comparison, it can be seen that the global public policies around AI are mainly moving toward two goals. On the one hand, promoting the development of AI technology can bring maximum benefits to the economy and society; on the other hand, we should try to minimize the costs and potential threats caused by AI to the economy and society. The two directions reflect two understandings about AI technology, that is, whether the use of AI technology focuses more on the creation of new value or only on the reduction of costs. Based on different emphases of policy objectives, two competitive AI policy trends have gradually formed. One kind mainly represented by the U.S. follows the principle of "permissionless innovation", which means tacitly giving priority to AI experiments in various fields and solving problems when they arise. The other kind, mainly exemplified by EU countries, adheres to the "precautionary principle", which means limiting or even prohibiting the application and development of AI in certain fields in advance according to the worst predication.

AI has brought economic benefits to public management. From a technical perspective, with the continuous expansion of the application scope of AI, the benefits it brings to the public sector are becoming increasingly prominent. From a practical perspective, in the context of the impact of the global financial crisis and the general fiscal deficits and insufficient administrative costs faced by governments around the world, introducing AI technology into public management is no longer a "should or not" problem but a "how to make better use of it" problem. On the whole, AI technology helps to break the cycle of mutual compromise between cost and quality; that is, it can not only reduce costs but also improve the work efficiency of government agencies, the job satisfaction of civil servants and the service quality of the public sector.

Specifically, the advantages of AI application in public management are largely manifested by cost reduction. Automation of administrative tasks can shorten working hours and reduce costs. Deloitte subdivided the government's package of tasks and made predictions through Monte Carlo simulation. The results showed that driven by high investment, AI technology will save 1.2 billion working hours for the U.S. federal government every year (27.86% of the total working hours of the federal government). If we convert these working times into monetary value, it can save $41.1 billion in labor costs. Even with low-level investment, the introduction of AI will save 96.7 million working hours (2.23% of the total), equivalent to $3.3 billion in labor costs, for the federal government every year.

In summary, AI technology has achieved remarkable success in reducing administrative costs and improving work efficiency and quality. As it proves to be more convenient, fast, reliable, widely used and low-cost in image processing, it has also been extensively used in object detection by public departments all over the world. Taking the U.S. postal system as an example, after bringing in computer vision recognition of handwritten envelopes, 25 million more letters can be classified each year, saving millions of dollars. AI technology has created unprecedented value in newer application fields, including unmanned driving and public health, such as tumor detection and screening for vascular sclerosis and other malignant diseases.

10.3 "New Order" for the International Community

As mentioned above, the progress of AI technology does provide chances for public sectors to realize the transformation of cost reduction and efficiency and quality improvement at the same time, and it also brings the possibility for countries to seize the competitive edge in advance. However, the development and wide application of AI technology have posed unprecedented threats and challenges mainly revolving around ethical issues, which have also increasingly attracted the attention of governments worldwide.

Therefore, some countries represented by the EU concentrate more on the risks that AI technology may pose and decide to apply it to limited fields with low risks. According to a survey of 979 technology pioneers, developers, business leaders, politicians, researchers and activists in these fields conducted by PEW in 2018, 37% of the respondents even believe that public life will not improve due to the development and application of AI technology by 2030, given the risks it brings. These concerns also hold true for the public sector and mainly focus on the following aspects.

First, it reduces the control over one's life. The public sector is prone to adopt cheaper and faster machine algorithms in its operation, and the decisions affecting the delivery of public services will gradually rely on code-driven tools. Therefore, human autonomy will face risks. Most ordinary people will more or less sacrifice their independence, privacy and the right of choice while enjoying the convenience of digital life. As the AI system becomes more complicated, apart from a few information technology groups who own and are responsible for the design and operation of the system, the blind dependence of ordinary people on the system keeps deepening, and then the average person gradually loses their understanding, selection and control of the decision-making process and results. As Karl Meter, the author of the book "*Computational Social Science in the Age of Big Data*", said, the welfare of people around the world in the future will depend on the "intelligent" decisions made by governments counting on AI and other technologies.

Second, data systems for efficiency and control are dangerous. Since the operation of AI technology basically depends on the sharing of personal information, preferences and other data and the technology itself will not incorporate human values and ethics into systems, some experts worry that data-based decision-making may be prone to serious logical errors, biases, and ethical and moral errors. If we put the problem aside, the future of AI will be shaped by those driven by profit motivation and power desire. Then, it will generate severe impacts on the morality, law and existing rules of human society. Justin Reich of MIT once said that in a capitalist society, the combination of humans and AI is to create a new way to enhance the monitoring and control of workers and then serve influential customers.

Third, the substitution of AI for human work may widen the economic gap and lead to economic and social turmoil. The general replacement effect and income effect of AI technology on employment have triggered much discussion. In the short term, the use of AI in the public sector is bound to eliminate a certain number of jobs. In the meantime, the corresponding new jobs needed are difficult to popularize on a large scale in a short time, and it is also difficult for the unemployed to transform from old occupations to new and more demanding posts soon, which will have a great impact on the economic income of the public originally engaged in simple work.

Fourth, the cognitive, social and survival skills of the public are declining. As humans' reliance on AI deepens, people's abilities to think and act independently and interact face-to-face with others have been impaired. More seriously, since various cognitive practices are transferred to machines, the public is facing the possibility of losing the virtues of judgment, empathy and love. In the long run, this harm will extend to the whole society. As Professor Daniel Siewiorek of the Human-Computer Interaction Institute of Carnegie Mellon University predicted, the negative effects of AI technology include human isolation, diversity reduction, loss of situational perception ability due to reliance on the GPS navigation system, etc.

Fifth, the public will become more vulnerable before excessive cyber crime and cyber war. Networked artificial intelligence is the combination of AI and networks. The development of networked AI is accompanied by increasing and sophisticated cybercrime as well as the possibility of creating panic in cyber warfare. Ordinary citizens who are likely to be exposed to cyber bullying, cyber crime and runaway cyber warfare because of disadvantages such as insufficient technical capability, information transparency, etc. Similarly, if the technology of the government public sector cannot be updated in time, it is also vulnerable to the damage caused by weaponized information. Their right of control may even be taken over by machines without knowing the true intent of some software, bringing irreversible damage to human society.

Facing the risks and challenges that arise from AI, we need to establish a new social system. The path followed by the current AI policy evolution is mainly based on objective fact analyses and subjective value judgment. If we only consider the technical facts when building the institutional system of AI, then the resulting social norms will have a severe lag. which will eventually lead to policy paralysis when faced with public propositions of social morality and ethics resulting from endless technology. Therefore, when we explore the construction of AI systems for the future, we should uphold "humanism" and build a perfect policy system, social regulation and risk control mechanisms from the aspects of safety, ethics and technology.

AI-related policies not only have the characteristics of general policies but also have special value content. The main reason is that AI needs to have general universal values, the necessary elements of which include personality justice, distribution justice and order justice, constituting the legitimacy basis of AI ethics. On the basis of these values, AI possesses special values in safety, innovation and harmony.

Security is the heart of AI policy and legislation and an important cornerstone of the harmony and stability of society as a whole. The security risks of AI mainly come from the possibility of AI surpassing human beings, the instability caused by the technical defects of AI itself, the unknowability of AI decision-making arising from black box algorithms, etc. With security as the core, on the one hand, it is necessary to ensure the preciseness and rationality of establishing rights and power of AI from the legislative level; on the other hand, legislation and policy also need to make full use of technical means to solve security barriers.

Innovation is the soul of the formulation of policies and legislation of AI. The specific background of the times often means that different laws and regulations uphold different values concepts and value emphases. Currently, relying on knowledge and skills to create new productivity has become the consensus of all mankind. Innovation has become the most typical characteristic of AI, and it needs to be reflected in all links of legislation and system design to realize the coordinated progress of top-level design and industrial distribution. The establishment of AI institutions that can promote innovation can be mainly achieved through the following three aspects. First, we need to promote the deployment and promotion of national development strategies with systematic thinking and uphold a holistic and global tone. The national strategy should actively guide and effectively propel AI and make good use of institutional advantages. Second, it is necessary to formulate policies and laws to boost and supervise AI from the industrial level, clarify access norms, set corresponding safety standards, and improve the supporting facilities of digital infrastructure to enable the legislative innovation of AI to possess technical, theoretical and industrial bases. Finally, we should strengthen the intellectual property protection and innovation incentive mechanism, focus on the development of a batch of enterprises with core technological competitiveness, and push ahead with technological innovation and the development of emerging industries. Strategic guidance and the consolidation of legal norms coupled with the implementation of policy incentives have expanded innovation from policy value to practical initiatives and then to constructive achievements.

Harmony and sustainability are the ultimate goals of AI policy and legislation. So-called sustainable development refers to adhering to the values of people orientation, benefit sharing for all, integrative development and scientific research innovation. We need to facilitate the ethical development and application of AI from the legislative level to ensure that AI values can complete the mission of benefiting mankind and realize the common progress of mankind.

Let us look at AI risk control dominated by law. The dignity of law lies in its execution, which requires close cooperation between supervision and standard-setting institutions. The law-based AI risk control system aims to build a comprehensive risk control mechanism composed of prevention in advance, in-process intervention and postevent disposal by establishing a complete regulatory standard system based on the theoretical basis of legal discussion. The guidelines below need to be followed.

1. **Reasonable application of prudential supervision can leave enough space for the growth of technology**

In the early stage of legal perfection, as AI technology is gradually penetrating all fields of the whole industry, many legal gaps or contradictions and conflicts between reality and legal provisions will emerge. Traditional laws have been unable to regulate and restrict various problems in the AI industry. In view of the current situation of the technology-driven new business modes and new demands, the State Council has put forward the principle of prudent supervision, which means to appropriately relax the policy and legal supervision in new fields, especially in technology with a prudent attitude to leave enough space for the growth of new technologies. As a link with the most technical strength and application value in new types of business, artificial intelligence needs more space in law for development and breakthroughs, rather than being firmly restricted by rules and regulations. However, the prudential supervision here does not mean full liberalization. For acts that challenge legal sanctity or are related to principles, the public security organs and judicial departments should implement necessary intervention measures in strict accordance with the provisions of the relevant laws. During the period of law making, perfection and transition, relevant authorities should accurately and dynamically control the limits of law enforcement, encourage bold technological innovation, and protect the legitimate rights and interests of the public, society and the state from damage.

2. **Regulatory technology and compliance guarantees can ensure that AI grows steadily on a reasonable track**

The purpose of regulatory technology and compliance guarantees is to balance the relationship between the development of AI technology and the protection of the legitimate rights and interests of the public, enterprises, the state and other relevant parties. Regulatory technology and compliance guarantee need to conform to the value orientation of taking society as the standard. Relevant regulators, standard design, system construction and legislative institutions should take responsibility. With the efforts in system, supervision and legislation, all related parties should leave sufficient development space for AI while adhering to the governance focus of taking data flow as an effective means to ensure that personal rights and interests are not infringed upon. In addition, through the establishment of rights with institutions as the main body, AI is guided to play an active role in social governance and public order maintenance in multiple ways, enabling every citizen to enjoy its achievements.

In terms of policy, the EU has paid a heavy price for "rigorous innovation" since it does not have one Internet company with a market value of 100 billion US dollars thus far. In contrast, China adopts a new economic support strategy of "giving AI reasonable and appropriate support". Facts have proven that innovation trial and error is the only way to revitalize business, especially in the next three years when the

10.3 "New Order" for the International Community

epidemic has dealt a heavy blow to the global economy. Therefore, we do not need to completely copy the regulatory policies of others but should focus on pushing ahead with relevant policy-making by encouraging innovation in an open-ended manner.

Building a unified regulatory framework is crucial to standardize the development of AI. Taking the European regulatory framework of the "ecosystem of trust") as a reference, we can first define risks at different levels, such as the risk of violating personal privacy, discrimination and other threats to basic rights, security risks and the allocation of responsibilities and obligations. Next, the existing legislation should be partially adjusted to make it more effective for artificial intelligence scenarios. Third, for high-risk scenarios such as national security, medical treatment, energy, etc., the regulatory framework should list the high-risk departments covered, review and modify them regularly, and clarify the significant risks that may generate in the use of AI in specific fields. Finally, the reliability of AI can be ensured from aspects including data training, data use, preservation and destruction, information transparency, system robustness, human subject status, etc.

Based on the basic principles of AI policy, we also need to establish some universal ideas. At present, the influence of AI on human security has made industry experts begin to question the credibility of AI technology, especially when AI is used in key industries such as education, medical treatment and biomedicine. We are wondering which direction each decision made by AI will directly guide human related rights. Whether product designers, technology users or ordinary users need to reach a universal consensus by setting AI technical ethics and making in-depth and systematic studies and judgments on the development of current ethics.

First, AI should follow the principle of human–machine mutual trust. AI products and services cannot rely on their strong abilities in unstructured data processing, computation, information collection and distribution to maliciously instill into human beings the content and values of deception, falsehood, terror and ethical standards against law and morality. AI technology should not be used to mislead or cheat for illegal benefits by virtue of the physical and psychological weaknesses of certain groups. AI is a kind of fully expanded human cognition. The mutual trust between humans and AI depends on the behavioral limits of the latter.

Second, humans and machines are equal and mutually beneficial in AI. Artificial intelligence should also not abuse the independent decision-making ability given by human beings, and all behavior standards of AI should conform to the human legal and moral codes of conduct. As AI is only a quasi decision maker without free will, human beings need to strictly limit the scope of their exercise of power and free decision-making to ensure that it can play an effective role within reasonable authority and human rights, development and freedom are fully guaranteed. AI should not carry out illegal monitoring, privacy data collection, threatening manipulation, etc., on individuals through technical advantages and monopoly, or infringe on the freedom and reputation or even the safety of life, health and property of others. The existence value of AI is based on equality and mutual benefit between humans and machines rather than AI dominating human beings.

Third, AI will coexist with human beings for a long time. Long-term symbiosis means the reduction of conflicts between them. As human beings, we should first fully understand the principles of AI products as well as the technical logic behind, comprehensively and objectively publicize the pros and cons of AI to the public and eliminate their unreasonable fear, discrimination and misunderstanding of AI. Second, AI developers should not intentionally or unintentionally implant their subjective feelings, especially their discrimination against culture, country, religion, class and gender, into the underlying structure of AI development. The research and development of AI should uphold the value of justice and goodness. Finally, man-machine symbiosis can not only effectively accelerate the upgrading of AI but also make human beings more proficient in dealing with interpersonal and human-machine relationships to realize the intelligent upgrades of human beings.

Fourth, human dignity and subject status are unshakable. The development of AI should always adhere to its status as a tool rather than alienating human beings into the tool. AI products and services should not exploit and oppress the mental and physical labor of human beings with their overall intelligent strengths. Meanwhile, they should not make human beings rely heavily on the virtue feelings they generate as well. AI growth should aim at further promoting human values and realizing human development in a free and all-round way.

The principles of AI development and governance are based on the above consensus. In 2019, the National Governance Committee for the New Generation Artificial Intelligence issued the *"Governance Principles for the New Generation Artificial Intelligence-Developing Responsible Artificial Intelligence"* and put forward a framework and action guide for AI governance.

Not only is the public sector keen on clarifying AI principles, but technology companies belonging to private sectors are also taking continuous actions in response to AI governance. Part of the reason is that people have great cognitive bias toward AI. The view that a lack of in-depth understanding of new technologies results in threat theory and large-scale employment has deepened public hostility and mistrust of AI technology. Problems such as privacy, responsibility, security and control, understanding and transparency, and discrimination brought by new technologies have increasingly aroused the interest of researchers. In the industry, in early 2017, the Future of Life Institute (FLI) put forward 23 principles for the development of AI-Asilomar AI Principles that aim to guide AI advances through a host of key propositions and precautions in scientific research, ethics and sustainable development, jointly ensuring the future ethics, interests and safety of mankind.

Technology companies in Silicon Valley are also expressing their views on AI public policies in multiple ways. For example, Microsoft CEO, Satya Nadella, proposed six principles for AI development in his speech in 2016. These guidelines include that AI must be transparent, pursue maximum efficiency without damaging human dignity, protect privacy, prevent bias, follow algorithm accountability, etc. All of them are the core design principles of the research and development of Microsoft.

Microsoft also issued the *"Artificial Intelligence Policy Recommendations"*. The document includes innovation laws and legal practices that aim to promote the development of AI, encourage the establishment of best practice ethics, measure privacy

10.3 "New Order" for the International Community

laws based on the benefits of AI, and recommend the government and the public sector to promote the spread and adoption of AI by launching major projects and systems.

Moreover, the Watson team of IBM set up an ethics review committee early on. IBM announced its three basic principles for AI growth at the 2017 World Economic Forum in Davos: not for the purpose of replacing human beings, increasing transparency and improving skill training and supply. Additionally, IBM wrote to the US Congress to state its public policy demands.

In the *"Artificial Intelligence Public Policy Opportunities"*, Intel conveys its response to the social impact of this new technology, including promoting innovation and open development, creating new employment opportunities and protecting human welfare, responsibly boosting data access, rethinking privacy issues such as PbD and FIPPS, ensuring ethical design and implementation and supporting accountability principles.

Google DeepMind's team recently announced the establishment of an AI ethics department, which shows that the company has put AI public policies such as ethics on the agenda and responsibly studied and deployed AI while strengthening the research and applications of AI technology.

Of course, the AI cooperative partnership jointly formed by five major technology giants (Apple joined later), such as Google and Facebook, also announced their emphases on the social impacts of AI, hoping that technology can benefit social and economic life and mankind.

Technology companies and industry organizations behind the technology industry are also on the move for this. On October 24, on behalf of the development interests and needs of tech industries such as Silicon Valley, the Information Technology Industry Council (ITI) of the United States issued the first *"AI Policy and Principle"*, which acknowledged that artificial intelligence, as a new technology, will exert revolutionary influences on social and economic life and productivity. AI systems can be used to solve some of the most urgent social problems, and they will create new job opportunities or empower workers instead of replacing them.

Generally, the top-level ethics represented by the principle of AI play a guiding role in AI development. AI should abide by human moral standards and ethics, adhere to people-orientation and be friendly to humankind. To promote the mutual understanding of human beings and machines at the ethical level, the first thing to do is to formalize the ethical norms of AI and embed moral standards and ethical reasoning rules into the underlying algorithm framework of AI as operators. Second, we need to establish a strong enough underlying algorithm encryption and active security defense mechanism from the technical level to strictly prevent illegal underlying falsification and data intrusion. In the face of malicious data falsification, algorithm deception and attack programs, AI can activate confrontation or defense systems to ensure the stable operation of systems and scientific and fair decision-making. Finally, access to underlying AI algorithms should be strictly restricted, and enterprises should also strengthen the technical ethics training for relevant developers and infiltrate the standardized consideration of morality and ethics into every link of AI education.

Finally, let us summarize a few thoughts on China's AI public policies. As a revolutionary technology, AI can create huge economic benefits, boost economic growth, and help solve a set of urgent social problems in human society, such as transportation, urban construction, medical treatment, and environmental protection. In the meantime, a good public policy environment and vigilance against risks and crises are also needed.

According to the *"Government Artificial Intelligence Readiness Index 2019"* released by Oxford Insight, a think tank, China ranks only 20th globally and 5th in the Asia-Pacific region. In addition to part of the missing statistical data, the main reason lies in China's capacities for scientific research and rapid transformation and application being inadequate. However, given China's rich data, the continuous growth in the number of AI engineers and the speed of launching AI applications, the think tank report also assumes that China has incomparable competitive advantages in the AI sector, and its government readiness index will catch up with or even surpass the top countries soon. In view of the above three strengths, China's AI policy can focus on the three aspects below.

First, China should encourage responsible AI research and development and build the country into a data powerhouse. On the one hand, we should open enough data applications to enhance the development of machine learning. On the other hand, we should also pay attention to the fairness, transparency and security of effective data. At the local government level, it is necessary to supervise whether the whole process of AI technology from development, design to promotion and application conforms to the objectives and usage specification of public management while making full use of it to deliver public services. The government needs to strengthen grassroots governance capacity while trying to avoid possible risks.

Second, we should attach more importance to education, talent training and labor force transformation. AI development requires the follow-up of education and the cultivation of high-end talent. As stated in ITI's AI policy principles, to ensure the employability of the future workforce, the public and private sectors should work together to design and provide work-based learning and training systems and actively provide students with practical working experience and specific skills. The civil servants available shall be trained to satisfy the demands of the development of AI technology along with the cultivation of new-type scientific and technological talent.

Third, an AI industry ecosystem should be built. To create a virtuous cycle of AI industry ecology, we need to promote the development, theoretical modeling, technological innovation, software and hardware upgrading and industrial implementation of a new generation of AI-related disciplines and enhance the support for the real economy empowered by AI, such as providing horizontal subject funds, tax incentives and other support for "AI+" projects. In addition, reform may hinder the development of AI and the applications of the current systems. However, radical legislation is not a good option. We should adopt a more flexible way of supervision that will not stifle the initiative and vitality of technological development. Overall, we should make concerted efforts in various aspects, including technology research and development, talent training, industrial application and policy support, to create a more innovative, efficient and intelligent AI industrial ecosystem.

Chapter 11
How Can "We" Realize Cogovernance in the Postepidemic Era

In the last chapter of the book, let us discuss the proposition of social governance in the intelligent era. Social governance is one of the most important areas we can see influencing the development of the smart age. Since there is no corresponding top-level design mechanism in China, all localities have formulated intelligent construction plans for social governance based on their own characteristics in hopes of becoming a model that can be copied and popularized. Shanghai Artificial Intelligence Pilot Zone, a project I participated in, is one of the examples. We should not stop at the research of pure "legalization of social governance". The "legalization of social governance may become empty talk if it is divorced from specific scenarios or social governance practices. In addition, we should also study the path of rule of law in specific social governance fields, such as intelligent social governance, and clarify the relevant scenes, which will help us understand the development path of social governance in China.

Additionally, research on social governance jurisprudence is rising in China. In recent years, various branches of law and its adjacent disciplines (such as sociology, politics, management, economics, etc.) have studied social governance and realized many research achievements. All these need to be systematically sorted out in the theoretical framework. The construction of social governance jurisprudence can effectively integrate the fragmented research results of various disciplines into thought-related and logically rigorous theoretical and academic systems.

Social governance jurisprudence is right a new interdisciplinary subject. It is a social science that takes jurisprudence as the theoretical basis, sociology, public management, politics and economics as the theoretical support, and social governance and its legalization as the main research objects. It is a theoretical knowledge system of interdisciplinary integration and practical application. Establishing a core category, determining theoretical propositions and optimizing research methods based on the discipline characteristics of the subject are the premises and foundations of constructing the "three systems" of social governance jurisprudence. The integrated construction of a "law-based society", law-based country and government is a major issue to be answered urgently in the process of expanding the governance path, system, culture and practice of socialism with Chinese characteristics in the new era.

In the last chapter, the author studies this interdisciplinary field and answers the key propositions on the intelligent development of China's social governance, which are intended to inspire more valuable thoughts. It is also a discussion on the value and application of the contents of the book in practice and an important starting point for our study on the discipline of social governance in the future.

Finally, what is worth mentioning is China's role in global governance. As COVID-19 is still spreading worldwide, many countries and regions are undergoing a large-scale pandemic, which makes China's future development gain more modernity, namely, uncertainty and instability. Considering that we are in a great change unseen in a century, the postepidemic world is more lacking global governance, and some systematic propositions are falling into crisis, we need to consider how to provide new ideas for global governance, that is, how to provide the public goods and governance logic in this smart era for global governance.

11.1 Public Goods from a Global Perspective

In fact, COVID-19 is the first real global crisis since World War Two. It is changing people's lifestyle and the pattern of global political and economic relations. Before this global disaster, deglobalization had already taken place. Since the subprime mortgage financial crisis in 2008, global support for economic globalization has been continuously weakened. The core logic lies in the development of the global economy bringing both winners and "losers". The latter mainly come from the middle and lower classes of developed countries. For example, the failure of de industrialization and redistribution of social wealth in the United States and other countries led to the rise of populism and conservatism, which caused numerous black swan incidents.

In terms of global development, there are three major uncertainties.

First, globally, when the epidemic can be under control is still unknown. When will China, the United States and other countries make progress in the development of vaccines and new drugs? To what extent does the continuous variation of the virus affect the actual efficiency?

Second, the actual situation of the global economy after being affected by COVID-19 is uncertain. Will it bring huge systemic risks in the economic field? Will it result in stagnation or even retrogression of the global economy?

Third, the impact of the pandemic on economic globalization and the international order is uncertain. To what extent will economic globalization be adjusted? What are the major changes and trends in global governance and international order? There are no clear answers to these questions at the moment. They can only be used as the basic judgment factors for us to understand and observe the impacts of the pandemic.

Let us return to a specific basic question. How should we provide public goods in the postepidemic era? In the last year's "Global Vaccine Summit", China expressed

11.1 Public Goods from a Global Perspective

its willingness to provide safe, effective and high-quality global public goods to the rest of the world, which may become an iconic turning point. The white paper titled "*Fighting COVID-19: China in Action*" issued in June 2020 presented us an image of a "responsible" global power. Whether from the perspective of the security of domestic people's livelihood or from the current global situation, China's strict anti-epidemic measures have shown a responsible attitude toward life. In regard to the delivery of public goods, China has displayed the idea of "cogovernance and sharing". How can we understand the concept of "global public goods"?

First, we should understand the concept of "public goods". In essence, public goods were originally an economic term that refers to the products and services provided by a government to meet the public needs of all members of society. Initially, the public goods theory was limited to one country only. The theory believed that the government had the responsibility to provide public goods, including national defense, diplomacy and public security, as well as social infrastructures, such as roads, bridges, road signs and lighthouses, to meet the development of the social economy and guide the optimal allocation of social resources. Within a country, the costs of public goods are raised by the government through taxation from the public.

During the Cold War, the system composed of the North Atlantic Treaty Organization (NATO) led by the United States and the Warsaw Treaty Organization headed by the Soviet Union was a typical international system of supplying international public goods through hegemony. International relations researchers have also sharply criticized the two critical defects of international systems while affirming the stability brought to the international community. On the one hand, international public goods have been "privatized" by hegemonic countries. Generally, "privatization" means that someone keeps public goods for his/her own and private use purposes. The "privatization" of international public goods here borrows this phenomenon, which means that the United States has turned the international public goods originally intended to serve the whole international community into its own profit-making tool. On the other hand, the phenomenon of "free ride" has caused a shortage of international public goods.

After the end of the Cold War, the collapse of the Soviet camp provided unparalleled opportunities and space for the all-around promotion of globalization. The internationalization of production, finance, information and technology showed an unprecedented momentum of development, which greatly liberated global productivity and freed various factors of production from the constraints of borders and boundaries. Thus, a trend of seeking optimization of resource allocation emerged globally. Driven by economic globalization and against the background of the pattern of "one superpower and multigreat powers" in the international political field, the abovementioned international public goods, which were originally limited to the Western camp, suddenly covered the whole world and were improperly exaggerated by some people as global public goods. For a period of time, the world trade organization was dubbed the "Economic United Nations", and the IMF was regarded as the guardian of the global economy. The World Bank, originally responsible for providing assistance for undeveloped countries, openly promoted the values of the

United States worldwide in the name of the Washington Consensus. The U.S. became the ruler of developing countries by linking economic assistance to human rights and national governance.

Therefore, although the WTO and the IMF were highly rated as the carriers of economic globalization, it was not difficult to find that the "privatization" tendency and undersupply of international public goods had obviously hindered the progress of economic globalization. First, people can see that after the end of the cold war, due to the relative decline of American power, the tendency of American "privatization" of international public goods turned out to be more obvious, which has become the main source of global system turbulence. In brief, for a long time in the past, the United States served as the main provider of global international public goods. The declination of its soft and hard power and the contrast between its status and actual strength resulted in a serious shortage of international public goods truly needed by the global society after the Cold War. To this day, we need new providers of public goods. They are not only in governments participating in the international order and international systems but also all sorts of affiliated nongovernmental organizations. For instance, the WHO is a kind of "global public good". The public goods provided actually determine the governance system adopted. Of course, the latter category is much broader than the former. Here is an example. The WHO is working as one of the main public goods providers of the global health governance system, while regional alliances (such as ASEAN, African Union, etc.) and other international organizations (World Bank, United Nations, etc.) also play an important part in global health governance.

Given that the original provider of international public goods, the United States, has withdrawn from a host of important global governance processes (such as the *"Paris Climate Agreement"*) and global governance institutions (such as UNESCO) and is likely to continue to reduce its contribution to "global public goods", a range of problems may occur. Therefore, here, we focus on the emerging providers of "global public goods", that is, the entry of emerging developing countries (such as China) and nonstate actors, which makes the international organizations of the current system more worthy of the name of "global public goods".

International relations scholar Joseph Nye once came up with a concept called "The Kindleberger Trap" originally coming from the economist Charles P. Kindleberger. In *"The World in Depression 1929–1939"*, he holds that those hegemonic countries with absolute superiority in politics, economy, military and technology can give play to their leadership and provide the international community with global public goods, including the international financial system, trade system, security system and aid system, to gain other countries' recognition of the international order they established. One of the vital reasons for the great depression in the 1930s is that there is no hegemonic state to provide global public goods such as an open trading system or serve as an international lender of last resort. With the excessive expansion of hegemonic countries or the emergence of domestic political and economic problems, both their subjective will and objective ability to provide global public goods were waning. If there is no emerging hegemonic state to assume leadership responsibility and continue this cause, it will destabilize international development and the economic and security system.

The view above was developed into "hegemonic stability theory" by Robert Gilpin; that is, international coordination and cooperation can be promoted only under the special conditions of the existence of hegemonic countries. In 2017, Joseph Nye proposed that if the dominant hegemons are neither willing nor able to provide necessary global public goods and so are the rising ones, then the resulting leadership vacuum in global governance will cause chaos in the global governance system and global security crisis. This is "The Kindleberger Trap". In other words, if the prime beneficiary of global public goods does not have the ability or willingness to exercise leadership and guide more resources into the supply of global public goods, it is impossible for other countries to do so. Because a number of actants are involved in this process, they cannot coordinate collective action. In this case, global governance will also fall into the tragedy of global commons.

To put it another way, globalization makes the "public bads" that originally existed in a country go global. The essence of global governance is how to manage and control "public bads" and provide global public goods through global coordination. Inge Kaul believes that if the beneficiary countries, groups and generations of a public product have strong universality, it is a global public product.

Effective global infectious disease prevention and control has the nature of global public goods. The UNDP has listed infectious disease control as one of the seven categories of global public goods. The United Nations "*Millennium Declaration*" also lists global health and security as one of the ten categories of global public goods. In addition, the secretariat of the International Task Force on Global Public Goods has classified infectious disease control as one of the six categories of global public goods. In the field of global health governance, the WHO is an intermediate global public good, while effective infectious disease control or health governance is the ultimate good. The performance of global health governance determines the supply condition of health public goods. Leading countries neither have the ability to provide global health public goods alone nor have the intention to push ahead with international cooperation and provide global health public goods by playing leadership in the global health governance mechanism. Then, the collective action dilemmas in global health governance will be unavoidable, and global health public goods such as infectious disease control will inevitably be in a serious shortage of supply, causing the deterioration of the global health and security crisis. This is the "The Kindleberger Trap" of global health governance.

In the COVID-19 crisis, although there is some global anti-epidemic cooperation, more often than not, countries prefer to fight the enemy separately due to the absence of global leadership. As a consequence, the global pandemic is nearly spiraling out of control. How to maintain the global health governance system, jointly provide global health public goods, and then cross the "The Kindleberger Trap" of global health governance have become tough challenges facing the international community, which is also the reason why we discuss "global public goods" today.

The above is our thinking and discussion of "global public goods" in the postepidemic era. On the one hand, with these opportunities and challenges, how to participate in and lead more global governance mechanisms through external cooperation mechanisms is an important issue for China. China is facing the historical choice of serving as a contributor or even one of the leading parties of "global public goods". On the other hand, the challenge of COVID-19 makes us aware that China also needs to arrange the corresponding resources and inputs based on its positioning and requirements and plan its strategy from the global perspective.

11.2 Governance and Innovation of Intelligent Society

After discussing global public goods, let us discuss the innovation of governance mechanisms after China enters an intelligent society. We have certain experience and strengths in this respect. First, we have experienced multiple emerging technologies, including the Internet, big data, and mobile Internet, at the technical basis and application level, summarized the corresponding experience and reached a consensus on accelerating the development of an intelligent society. In addition, in terms of research methodology, we are pushing for the expansion and innovation of relevant research, such as social governance jurisprudence and other fields. Therefore, in this part, we will introduce our work on governance mechanism innovation from these two angles, providing a new perspective for us to understand ethics and governance in the intelligent era.

From past experience, we have two methodological advantages in constructing an intelligent social governance system, and one of them is system integration. Big data technology can help us understand and grasp social contradictions and progress in general and achieve governance efficiency of "$1 + 1 > 2$" in multiple dimensions. The other is deep learning. Intelligent core driving forces should be built by data-intensive brain computing methods to realize scientific and refined social governance. With the rapid development of socialized mass production, the geographical and industrial boundaries are gradually broken by the large-scale flow of population and production factors, and the nonlinear dynamic characteristics of social cyberspace are becoming increasingly distinct. However, facing the changes in the overall social picture under the background of globalization, the systematic effect of the social governance system has not been brought into full play, and problems such as fragmentation and decentralization in the social governance system have not yet been fundamentally solved. The intervention of intelligent technology will greatly improve the integrity and synergy of the social governance system, strengthen predictive and forewarning abilities to prevent various risks, and then realize sustainable and effective long-term management that can address symptoms and root causes.

The pandemic since 2020 is actually a test of the level of governance and intelligence of our society. China has made great achievements and gained some experience in this field. The intelligence of social governance in the era of big data highlights the technological leading role of governance methods. With the overall deployment

11.2 Governance and Innovation of Intelligent Society

of 5G, artificial intelligence, industrial Internet and other new infrastructures, related services around data collection and information sorting, such as big data centers and cloud computing centers, will flourish. The widespread application of big data technology in strengthening and innovating social governance will deepen our understanding of the operation and governance rules of intelligent society. In this context, the intelligence of the social governance system should be actively boosted and the efficiency of social governance should be improved by virtue of advanced ideas, scientific attitudes, professional methods and fine standards to help enhance the predictability, accuracy and efficiency of social governance.

A typical example is the implementation of China's new infrastructure policy this year, which shows the significance of the database of social governance. Recently, the new infrastructure has been widely valued. It is a technological innovation-driven and information network-based infrastructure system guided by the new development concept. The system is designed to meet the requirements of high-quality development by providing services such as digital transformation, intelligent upgrading, integrated innovation, etc. TThe new infrastructure will significantly expedite the digital upgrading of the social governance system. Benefiting from the current industrial digitization and digital industrialization, social governance systems have the material conditions for big data collection and calculation.

The new generation of mobile communication technology will drive society into the era of the Internet of things. The deep integration of various domains, including 5G, cloud computing, the Internet of Things, and artificial intelligence, has promoted the formation of the core competence of a new generation of information infrastructure. In particular, the 5G network features high transmission, low delay and extensive connection compared with 4G. The application scenarios involve enhanced Internet, 3D video, cloud office, augmented reality, automatic driving, smart city, smart home, etc. The new generation of "information superhighways" with 5G networks as the main structure will provide a wide high-speed transmission channel for large amounts of data and information, which guarantees the powerful information capability of the social governance system.

In addition, city brain technology has become a critical infrastructure of social governance. Since the socialist social governance system with Chinese characteristics emphasizes integrity and synergy, under the condition of the market economy, the social structure presents the characteristics of diversification and dispersion, bringing challenges to collaborative governance. The intelligence of the social governance system is supported by AI technology. Artificial intelligence is something like a cloud brain. It relies on the data from the "information superhighway" for deep learning and system evolution to complete machine intelligence. Currently, the city brain construction in the ascendant is a typical representative of this process.

In terms of the current practice, the city brain is mainly a digital interface created for public life, including the application scenarios of several systems, such as transportation, digital tourism, health, and emergency flood control. It generates hundreds of millions of collaborative data every day. The online data from all resources are not "static" but "dynamic". All theme scenes gathered by the city brain are in the present continuous tense. As the granularity of data collection becomes increasingly finer, the city brain can efficiently and conveniently grasp the exact information and event data of social governance scenes. Additionally, in the construction of the city brain, various networks are effectively connected, and the coordinated operation of information access and access equipment has broken the data islands among departments, enterprises and groups, thus promoting the establishment of a three-dimensional and networked social governance system.

To realize the aggregation of social governance data, localities have focused on driving three tasks in the practice of intelligent construction of social governance.

First, we need to achieve data convergence. The practice of social governance produces data from different departments and levels, as well as the personal data of enterprises and citizens. However, the data owned by a department, an enterprise or an individual cannot form big data, nor can it support the requirements of intelligent construction of social governance. The data of social governance must be combined through technical means and certain working mechanisms to form big data to realize the development and utilization of big data by different subjects based on their actual needs and build their own intelligent application systems by learning modeling with a full amount of data from different dimensions of social governance.

Second, we need to realize data integration. The social governance data collected by different subjects do not completely follow the same technical standard and may also be stored in diverse systems. Therefore, they may not be able to form big data even if combined into a big data collection. To this end, it is necessary to build a unified system or data standard to realize compatibility among data from different systems.

Third, it is necessary to build a unified Internet data platform. In the age of intelligent technology, data themselves appear in the form of raw materials. What is shared by departments, governments, enterprises and the whole society is the data resource itself, rather than finished product services. To store data from all aspects, we need to build a unified data platform. At present, big data can be gathered mainly through cloud platforms, and then realize the development and application of intelligent technology in big data with cloud computing serving as its analysis tool.

The above is our discussion of the basic concept and category of social governance as well as the reform of data elements. We have also explored the basic contents of social governance jurisprudence and its possible categories and characteristics, which has laid a foundation for us to understand the legal path of social governance and for the future direction of jurisprudence.

After understanding the database of intelligent social governance, let us discuss its legal path and problems relating to intelligence.

First, we need to discuss general problems of social governance and the rule of law, namely, the value of social governance. The consensus of modern civilization is that the key to governing a country and a society is to set, stress and abide by rules. Law is the largest and most important rule of governance. The rule of law is the best model of human social governance. To stimulate social governance, we must adhere to the rule of law. Therefore, social governance is the basic direction of national governance modernization. Good law and good governance and multigovernance are the two basic principles of the modernization of national governance. The maximum autonomy between nations and the public and cogovernance within the scope of the law need to be achieved.

In short, "intelligence" is based on big data and artificial intelligence. Therefore, improving the intelligence of social governance often means making full use of big data technology, cloud computing, the Internet of Things and other intelligent technologies in the process of social governance and realizing the deep integration of intelligent technology and social governance. In doing so, the modernization level of social governance can be enhanced comprehensively.

Achieving the governance of an intelligent society requires not only the empirical results mentioned above but also theoretical breakthroughs. We see that cogovernance in social governance is of great significance, and we also mentioned the differences between social governance jurisprudence and traditional jurisprudence research. Then, let us discuss the research of social governance jurisprudence.

Social governance jurisprudence is a multidisciplinary knowledge system with three core concepts: law, rule of law and jurisprudence. The whole legal theory system, together with the knowledge system and discourse system of law, are based on these three concepts. Legal issues and legal discourse can be found in national governance, social governance and the public sphere, and they are everywhere. It can be said that the category of jurisprudence is contained in the legal ideological system, legal system and legal operation system. It appears almost everywhere and at every time, and it seems to be omnipotent. Social governance jurisprudence should not only study social governance law and governance according to the law but also study the deep-seated legal issues of social governance modernization and social modernization to refine the legal concepts, propositions and discourse of social governance, build a scientific theoretical system, and integrate social governance jurisprudence into the trend of the new era of Chinese jurisprudence.

To do a good job in the research of social rule of law, at least three aspects need to be concentrated on. The first aspect is the background and conditions of the discipline construction of social governance jurisprudence. Social governance jurisprudence based on the demands for compound, innovative and capable high-level social governance talent should respond to the double first-class policy put forward by the central government and develop a new growth point for the development of the jurisprudence discipline. The second is the nature and structure of social governance jurisprudence. The orientation of the discipline is an interdisciplinary secondary discipline of comprehensive jurisprudence. The research content of social governance jurisprudence comprises two parts: the basic theories of social governance jurisprudence and the implementation of social governance jurisprudence. The discipline system of social governance jurisprudence covers the theoretical system, teaching material system, curriculum system, tutor team system and talent training system. The last aspect is the path of the discipline construction of social governance jurisprudence. The construction of this discipline is a combination of the teaching staff under the mechanism of external introduction and internal training, the international and open talent exchange mechanism, the problem-oriented assessment and evaluation mode and the construction of a collaborative innovation platform.

In fact, we can see that the emergence of social jurisprudence responds to the trend of world jurisprudence. In recent years, legal assessment taking legal indicators as standards has been carried out globally. According to the idea of the socialist legal system with Chinese characteristics, Chinese legal experts have proposed that China's legal indicator system should contain six first-class indicators: legal norms system, legal implementation system, legal supervision system, legal guarantee system, innerparty regulations system and legal effect indicators. The structure of the legal system is the dividing evidence of the first five indicator systems. In regard to the functions and effects of the legal system and the goals to be achieved, the legal governance effect needs attention. The effect indicators of the rule of law include power control indicators, human rights indicators, order and security indicators and the concept of rule of law.

Next, we will discuss the relationship between technological progress and the rule of law arising from the advent of an intelligent society from three aspects.

(1) Information technology has brought profound transformations to property attributes and property concepts. The new technological revolution has brought changes to social life and social relations, generating new interest relations, deriving new interest conflicts and triggering new interest disputes. The Chinese law community should extract rules to resolve conflicts of interest. As long as the technology is specific and independent and has the attributes of property, we should break through the limitation of the concept of real right. So long as the rights related to technology are legal and specific right type, they can be confirmed and publicized through registration and other information approaches. Similarly, rights can be independent of other rights and reflect their peculiarities. Taking virtual property as an example, it is the right of which can be disposed in law as the specific right of the subject.

(2) Accelerating the transformation of the rule of law is a vital response in the intelligent era. Legislative, judicial and legal research should respond to the impacts and challenges of the intelligent era. We can see that the impacts of the intelligent era on the rule of law are mainly reflected in three aspects. First, the intelligent society has brought a certain impact on the traditional law adjusting the social relations in the real society. In criminal law, traditional crimes are real space crimes, but with the advent of an intelligent society, crimes in cyberspace should also be adjusted by law. Since traditional law cannot adjust new illegal acts in cyberspace, legal shortages bring new legislative needs. Legislation should respond in a timely manner to various social issues brought about by the development of an intelligent society. The second problem is that the emergence of intelligent technology has brought a significant impact on judicial activities, such as online dispute resolution, evidence collection and fixation of network crimes. Therefore, judicial activities also need to comply with the progress of an intelligent society. The third problem is that the era of intelligence has put forward new topics for legal study or research. For instance, after virtual property emerges, determining the nature of the criminal acts of virtual property

violation involves not only the common concern of criminal law and civil law but also the coordination between legislation and judicature.

(3) Technological progress and innovation are important to economic and social development. However, technological progress not restricted by rules will also cause public concern and panic. The risk society has already come. Practice has proven that an industry that relies too much on self-regulation can also fail, while excessive or immature regulation will inhibit innovation. Innovation in the form of a sharing economy can be reflected not only in technology and economy but also in society. The goal of legal regulation is to encourage innovation and pay attention to public safety rather than inhibit development. Legal regulation should not limit the development of the sharing economy in China but should become an important institutional basis for its development and growth in China. The preset orientation of the human nature of administrative law in risk provides the most basic guidance for administrative law to effectively respond to the governance problems in a risk society.

In conclusion, since we are in the early stage of an intelligent society, we need to understand the basic framework and logic of social governance from the perspective of the rule of law. We will conduct studies on AI-related ethics, governance and rule of law from three basic perspectives.

First, we need to be problem-oriented and keep an open mind to conduct research.

Second, we need to carry out research in corresponding fields with the goal of perfecting China's jurisprudence discipline system, especially focusing on the construction of emerging disciplines and interdisciplinary disciplines. The research of jurisprudence needs the support of other disciplines.

Third, we are conducting research on the basis of the theoretical consensus we have reached on issues related to technological development, social governance and the rule of law. This consensus includes several aspects. We should promote technological development, economic construction and social governance on the track of the rule of law. Governance according to the law and social governance are symbiotic, interactive and complementary. The legalization of social governance is the concrete practice of the construction of a country ruled by law in the field of social governance. The legalization of social governance needs to respect social autonomy. Accelerating the construction of the system of jurisprudence with Chinese characteristics requires the combination of Chinese experience and international view. Social governance emphasizes the joint participation of multiple subjects and equal consultation.

11.3 Truth of AI and Reflection of Black Mirror

In the last section, let us look back in history and explore the truth of the world, discuss the core of intelligence and the logical changes of human civilization in the history of computing, and finally obtain some thoughts on the ethical development of artificial intelligence from the challenges of the real world as the end of the body content of the book. It should be noted that only problems and challenges are innumerable in this age, and what we need is a broader vision and pattern when looking at them.

First, we will discuss the category of intelligence so that we can determine in what sense computers or robots have intelligence.

Through the study of animals done by Frans de Waal, an American Dutch animal behaviorist and primatologist, he tries to express a view that animals do not learn only through conditioned reflex and many animals also have cognitive abilities like us human beings, although not exactly the same as ours in his book "*Are We smart Enough to Know How Smart Animals Are*". Animals can not only make and use tools but also participate in various games and form certain intelligent behaviors. In other words, people and animals are continuous in cognition, but this is not to say that the intelligent arrangement of each animal is a neat sequence. Actually, it is more like a polyhedron. Each animal has its own unique cognition. Thus, from the aspect of intelligence, humans are unique animals but are not better and better than other living things. We need to think about the concept of AI from this aspect. Organisms have developed different intelligences to survive, and AI will also develop its own category of intelligence based on its survival law.

Based on the above ideas, let us discuss the governance paradigm in the practical sense of AI as a "beneficial machine" and understand the moral value of this round of artificial intelligence. This is not just because this round of AI originated from the mathematical basis of statistical learning but because almost everyone is paying close attention to the changes of relevant data to obtain the cognition of the impacts of the whole pandemic since it broke out.

William Petty calls statistics "political arithmetic". This is actually an example of how statistics can be used as a governance tool in areas such as sociology. In contrast, the morality of current AI technology is also reflected through the statistical "algorithm", and its core value of governance lies in whether it can find a stable order in a large amount of data.

This concept seems self-evident to statisticians in the 19th century. They believed that social data contain social order, natural data contain natural order, and statistical thinking is the only way to find a unified order. To put it another way, from the perspective of governance, the core of today's AI technology is to realize the core of enlightenment—the unity of natural order and social order through accurate algorithm governance (data thinking) and scientific technical governance (implementation mode). After finding this fulcrum, we can understand how to realize that AI ethics and governance actually reflect the positive values and ethical norms of society in statistics and then practice them through algorithms and technologies.

Finally, let us look at the current world landscape and the impacts of AI technology on our future. Here, we use "reflection in the dark mirror" to think about this problem. The reflection here implies that we think that AI is the reflection mirror of human social ethics. The understanding of AI is actually based on our understanding of the general map of human society and human civilization. Generally, there are three dimensions to consider.

First, AI technology is the mirror of human ideas and emotions. The new round of development of AI technology has not only brought about the rapid development of industry but also gradually become a new kind of philosophy and world outlook. When we take about AI technology, we end up discussing humanity and society, and all views will ultimately fall on human beings. For instance, a large amount of AI art is springing up. Artist Vladan Joler displayed his work *"Anatomy of an AI System"* in the "40 Years of Humanizing Technology" exhibition. He showed the labor, data and planetary resources required by an Amazon Echo sound in a visualized manner to remind us that artificial intelligence, as a technology, is also consuming resources through materialized objects, and its production process is not pending but coexists with labor, resource consumption and sustainable development. From this level, AI is the embodiment of our ideas and emotions.

Second, AI technology is the mirror of human history and civilization. In his book *"Leviathan"*, Thomas Hobbes boldly conceived that people can create all sorts of artificial lives as well as a rational society leviathan through technology. The prediction has become a reality with the development of technology. It can be seen that one of the technological prospects of AI is to make human evolution rely more on technology rather than biology and form a new trend of civilization along the evolution of cybernetics, systematology and relational realism. As Bernard Stiegler said, human beings are artificial and technical; that is, human beings cannot find meaning in themselves but need to find it in the artificial limbs they make and invent, which means they are doomed to drift when free. He calls it "original disorientation". They want to invent their others and their existence, he stated. The statement has profoundly pointed out the relationship between human beings and AI. We understand the world through a technical paradigm and create new extensional connotations of human beings at the same time.

Third, AI technology is the evolution mirror of the business and technological innovation paradigm. In my opinion, when we discuss the governance of artificial intelligence, it not only contains the governance of the negative impact on AI technology but also has greater connotation and extension. If we hold that the target of promoting AI is to improve the overall well-being of society, how to stimulate it to solve the problems in the social and economic fields of human development, promote the healthy development of society and improve practical efficiency is the core of our governance promotion. Hence, both the "sustainable development of AI" we are advocating and the "justice innovation" mentioned by some domestic scholars are viewed based on this logic. Promotion of the upgrading of all innovation elements by distinguishing the paradigm of the "Schumpeter innovation model". In other words, enough social elements need to be incorporated into innovation apart from market elements to promote the implementation of the social value of AI, which is our recognition of innovation in technology and business as a whole.

The content above is our discussion on the proposition of artificial intelligence governance, and the research work on this part is still being advanced. How to drive the common governance of society, the sustainable development of AI and the implementation of a more systematic innovation mechanism of AI are our major research directions. We also hope that this chapter will provide you with new thoughts on how to understand AI ethics and governance in the postepidemic era.

In this challenging era of intelligence, in the face of the proliferation of cognitive capitalism, the blurring boundary between human values and knowledge, and the identity problems caused by AI labeling, it is time for us to rerecognize AI, reunderstand the relationship between humans and AI and reread Marx's "Fragment on Machines". Marx stated that the proliferation of fixed capital shows to what extent general social knowledge has become a direct productive force and to what extent the conditions of the process of social life are controlled by the general intellect and transformed based on this intellect.

We find that when Marx said that social life is controlled by the "general intellect" and transformed according to this intellect, what he emphasized is obviously not "individual intelligence" but the overall intelligence of society. However, the future development of AI should concentrate more on how to promote the overall intelligence of human society to raise up to more challenges.

The author believes that technological mutations will not only subvert business innovation but also increase the general happiness and dignity of mankind to wholly enhance our expectations and confidence in the future.